Founders of Modern Mathematics

Founders of Modern Mathematics

F. Gareth Ashurst

FREDERICK MULLER LIMITED
LONDON

First published in Great Britain in 1982 by
Frederick Muller Limited, London SW19 7JU

Copyright © 1982 F. Gareth Ashurst

All rights reserved. No part of this publication may be reproduced, stored in a retrieval system, or transmitted, in any form or by any means, electronic, mechanical, photocopying, recording or otherwise, without the prior consent of Frederick Muller Limited.

British Library Cataloguing in Publication Data

Ashurst, F. Gareth
   Founders of modern mathematics.
   1. Mathematics—History
   I. Title
   510'.9   QA21

ISBN 0-584-10380-8

Typeset by Texet, Leighton Buzzard, Bedfordshire
Printed in Great Britain by Redwood Burn Ltd.,
Trowbridge, Wiltshire

Contents

Acknowledgements

List of Illustrations

Introduction

1	Evariste Galois	1
2	Sir William Rowan Hamilton	15
3	George Boole	29
4	Arthur Cayley	45
5	Richard Dedekind	55
6	Georg Cantor	69
7	Felix Klein	83
8	Giuseppe Peano	99
9	David Hilbert	115
10	"Nicholas Bourbaki"	127
	Postscript	134
	Further Reading	138
	Appendix: Group Axioms	140
	Index	141

Acknowledgements

THE WRITING OF THIS BOOK would have been impossible without the use of numerous sources of reference. These include many learned articles, obituary notices and the works of the mathematicians themselves, which are scattered throughout the literature of mathematics and are not directly available outside the large academic libraries. The extensive biographies of Galois, Cantor and Hilbert by Leopold Infeld, J. W. Dauben and Constance Reid (full details of which are given in the section on Further Reading) have been consulted for the respective chapters. Likewise, I have used the group biographies of E. T. Bell and Alexander MacFarlane and the general histories of mathematics by Morris Kline and Carl B. Boyer.

I would like to add my very special thanks to Professor Hubert C. Kennedy, who was kind enough to send me offprints of a number of his articles on Peano. As well as being one of the world's leading authorities on the life and work of Peano, he is almost the only one who writes in English. Through my use of his articles and two books the chapter on Peano owes much to the work of Professor Kennedy.

Even so, all the mistakes and inaccuracies which there are in this book are mine and mine alone.

I would like to add my thanks to my wife Shirley for her moral support and critical and secretarial help.

List of Illustrations

Evariste Galois	facing page 1
Sir William Rowan Hamilton	14
George Boole	30
Arthur Cayley	47
Richard Dedekind	56
Georg Cantor	70
Felix Klein	84
Klein bottle	84
David Hilbert	114
Kurt Gödel	137

Introduction

THE HISTORY OF MATHEMATICS, like any other history, is made by people. The evolution of new ideas or the development and combination of old ones is so dependent on the opportunities and personalities of the people involved that the history of mathematics cannot be separated from the lives of the mathematicians who created it. Among the famous mathematicians there are revolutionaries, statesmen, prodigies and phantoms. Rarely are they as dull as is commonly thought, and frequently their lives are as interesting as those of warriors and explorers.

Indeed, mathematicians are explorers of human thought and accomplishment. Their achievements come about as the result of new ideas or, more frequently the fortuitous combination of old ones which have lain hidden in embryonic form waiting for a suitable set of circumstances. At the start of the nineteenth century mathematics was, as at almost any other time, progressing along well established lines. Subjects such as calculus and its extensions in differential equations, algebra, geometry and the theory of numbers were following well established paths. However, innovations were always being sought and new areas of study developed. In these new areas some dormant ideas came into their own, along with some very new ones, to the extent that significant and often revolutionary changes occurred. Some of these new notions and concepts were of such importance in themselves that they influenced much more of the subject than their initial area of importance. The resulting developments have often been grouped together under the title of 'modern mathematics'. This is a misleading description since it has little to do with time. It refers to those parts of mathematics which have risen from small beginnings in the last century to positions of paramount importance in the present one. Because of this importance they frequently take up a large part of school, college and university courses. It is the men behind these developments who were the founders of modern mathematics.

Evariste Galois (1811-1832) *Roger Viollet*

CHAPTER ONE

Evariste Galois

EVARISTE GALOIS lived during the turbulent times in France at the beginning of the nineteenth century. As a Frenchman he was very much a man of his time and led a very exciting life as a republican revolutionary; but as a mathematician he was very much ahead of his time.

He was born on the 25th October, 1811, in the small town of Bourg-la-Reine, a short distance from Paris. His father, Nicholas Gabriel Galois, was director of a small boarding school and the local mayor. He had been elected to this office during the Hundred Days after Napoleon's escape from Elba and, although he was an ardent liberal, he kept this office during the numerous changes in French politics and served faithfully under the monarchy. Evariste's mother, Adelaide-Marie Demonte Galois was a very intellectual and somewhat aloof person who came from a family of influential lawyers.

His early childhood was very happy but his father led a very busy life running his school and performing his civic duties. Both his mother and father were of an intellectual disposition with interests in literature, the classics and philosophy. The family was very close but most of the companionship and warmth the young Evariste received was from his elder sister Nathalie-Theodore. All of his early education came from his mother who taught him Latin and Greek as well as the normal things a young man would need to know.

Evariste was very clever and learned quickly, so much so that when he was twelve he was able to go to a Royal School, the lycee of Louis-Le-Grand in Paris. In the school many of the cleverest people in France had and were to receive their education. This meant that the standards were high and the pupils generally very talented. Because of this it is strange that this and other such schools in France were rigid and inflexible. It was a boarding school and the life was completely regimented and governed by a strict timetable. Poor sanitation and meals conspired with dry teaching to either deter originality on the part of the students, or to make the life of a person inspired by different and exciting ideas unbearable.

Galois' school life must have been unbearable. According to his school reports he frequently daydreamed and chattered and his work was what might be expected of such a person in this choking environment. It was often hastily done without sufficient care or attention. He must have been well ahead of many of his fellows when he entered the school, but those who were prepared to conform to the narrow but considerable demands of their teachers fared better than he. He must have spent many hours longing for escape.

This escape nearly came soon after he joined the school. At this time the priests of the Society of Jesus, or Jesuits, held many offices and positions and wielded considerable political influence. Their influence was so strong that many minor officials did their best to ingratiate themselves with this priestly order and were known not very affectionately as Jesuits of the short coat. The school of Louis-le-Grand had once been a Jesuit school, and it was rumoured that it was to be returned to their authority. A rebellion against this was planned by the students but was discovered and the ring leaders expelled. The day following this expulsion was Saint Charlemagne's day, and the rather dictatorial director, M. Berthod, assembled some of the senior scholars to celebrate the proceedings. These included drinking a toast to Louis XVIII. Either by grand design or by simple reluctance, the pupils did not perform this mark of respect for their king. As a result about one hundred and fifty of the most able scholars in the school, and possibly all of France, were expelled. Evariste missed this because he was not part of the gathering, but the memory of it lingered with him for the rest of his life.

When Galois entered the school he went into the fourth class. Over the years he progressed to the second class, and was about to enter rhetoric or senior class when it was felt that he would benefit by repeating the second class, especially since he was quite young compared with many of his companions. Although his idle and daydreaming ways had caused this, it was very much the means of his escape from the regulated oppression which had caused him so much anxiety and loneliness. To make a change from having to repeat exactly the same course he entered the mathematics class; mathematics was not thought important enough to be compulsory. The mathematics that Evariste knew at the start must have been very elementary indeed but he found the new work easy, exciting and logical. The class textbook on geometry was *Elements de Geometre* by the great Legendre which in present terms would look very formal, but Galois understood each sentence, theorem, construction and problem immediately. His teacher M. Vernier was, like most of his teachers, an uninspiring person on the whole, but he did recognise that Galois had some talent and was quick to learn his subject. Galois' interest was fired so much that he got a copy of the *Resolution des equations numeriques* of Lagrange from the school library. In this algebra textbook Evariste discovered that the orderliness which he found in geometry was far in advance of that of algebra. Algebra was something of a hotch-potch of various ideas and methods with many unsolved problems and difficult areas. These problems were to occupy his spare time, and even his lesson time, from now on.

One idea that was to fascinate him and eventually bring him fame was concerned with the solution of equations. We know how to solve quadratic equations: we can factorise those such as $x^2 - 5x + 6 = 0$, or we can in general solve any equation of the form $ax^2 + bx + c = 0$ by the formula

$$x = \frac{-b \pm \sqrt{b^2 - 4ac}}{2a}$$

Also formulae are, and were in Galois' time, known that could be used to solve cubic equations of the form $ax^3 + bx^2 + cx + d = 0$, and quartic equations which are similar but contain terms in x^4. The formulae for cubic and quartic equations are much more complicated than the quadratic formula.

However, when we try to solve quintic equations, or equations of the fifth degree (containing x^5), things begin to go wrong. There is no formula, not even a very complicated one, which can be used for every quintic equation, even though many special equations can be solved fairly simply. Also any attempt to find one seems to result in a more complicated equation than the one which is to be solved.

Galois must have pondered these facts for many hours but after a while he began to make progress and perform real research in mathematics. This occupied most of his time and also stimulated his other work. That year he took the second prizes in both mathematics and Greek. He now had a goal. He knew, at least in part, what he wished to do with his life: he wanted to become a mathematician. The best way to accomplish this was to enter the Ecole Polytechnique, a college originally intended to educate army officers in engineering and science, but now the most prestigious scientific institute in France. To gain entry, it was necessary to pass an oral examination. Unfortunately the staff members at the Polytechnique were from the same mould as those at Louis-le-Grand. Evariste's lack of elaborate explanation and his less than punctilious devotion to the more obvious points, which were mere details to him, of mathematics let him down. One of the great mathematicians France has produced failed his examination in mathematics to enter the Ecole Polytechnique. Only one excuse can be given for his examiner; that had Galois mentioned any of his own work, he would, through his own ignorance, have found it incomprehensible.

The following year Galois entered the senior class and continued his mathematical studies with M. Richard. This new teacher was a mathematician of ability, with a general interest in the subject and considerable insight where both mathematics and Galois were concerned. He recognised Galois' talent and encouraged him to develop it and guided his reading and research. Richard showed him the work in the same area that had been carried out and published by another young mathematician, Niels Henrik Abel. There had been a time when Galois believed that he had found a way to solve the quintic, or general equation of the fifth degree, but he later found his error. He was pleased to learn that Abel had at first believed the same but had later produced a proof that such a solution by radicals, using the common operations of algebra and the extraction of roots, was impossible. Richard guided Galois in the preparation of his first published research on periodic continued fractions, such as

$$1 + \cfrac{1}{2 + \cfrac{1}{1 + \cfrac{1}{2 + \ldots}}}$$

By now it was 1829 and Galois was seventeen years old. His life had not been particularly happy so far but his way ahead seemed clear and likely to bring satisfaction and possibly happiness. However, it was from this time that a particularly disastrous and cruel sequence of events began.

The paper which Galois had been able to publish on continued fractions was basically elementary and was well understood by its readers. It was a good piece of work by a talented, budding mathematician but it suggested nothing of the new and important discoveries that Galois was making. The next paper he wrote was concerned with his discoveries on the solution of equations. It was difficult to understand mathematically, and as well as being an entirely new approach to the subject, it was written in a very condensed form. He sent it to the Academy of Sciences for publication. Normally, before a new piece of work is published it is sent to an expert to make sure that it is important, correct and generally worthy of bringing to the attention of professional mathematicians. The referee to whom it was sent was one of the leading mathematicians of France, Augustin-Louis Cauchy. Cauchy was not a young man and he was extremely busy with his own research. He received many manuscripts such as Galois', most of which were more easy to understand. In all events, the paper did not return from Cauchy. When Galois made enquiries about its fate several months later, it was discovered to have been mislaid; it has never been found.

Had the true worth of this paper come to the notice of the French mathematicians, Galois might have been saved from the next irony of fate to which he was subjected. He made his second attempt at the examination for entry to the Ecole Polytechnique.

The examiner was Dinet who, much earlier, had been Cauchy's teacher. Dinet had supervised the examination interviews for nearly forty years and had met students of all shades of ability, yet he expected students to give stereotyped answers to his questions in line with French school mathematics. Galois gave deep but briefly explained answers. Dinet asked what Galois knew about logarithms, no doubt expecting to be told almost sentence for sentence the definition and rules for these as they appeared in the popular textbook by Leonhard Euler. Galois did not oblige with such an easily digestible answer. He launched into his own explanation on the subject. Dinet neither expected his answer not seemed to understand it. He asked for greater elaboration and Galois tried again but was still misunderstood. The examiner was not led back to the familiar ground, but deeper into the subject. He still did not understand. Galois became nervous and irritated at the same time. The old man was ignorant and stupid and could not understand. Such ability at simplification as Galois had was lost in his rising anger. He tried to write explanations on the blackboard, but they still did not suffice. Finally his temper, or sense of futility on realising that all was lost, got the upper hand and he threw the blackboard rubber at the man whose ignorance stood between him and his aspirations. The throw was a success, but the interview was a failure. Galois had suffered the fate once again, of a man ahead of his time.

His missed last chance of becoming a professional mathematician was not to prove the worst that was to happen to him during the remainder of his stay at school. His father had survived the constantly changing French political climate well. He was still the liberal mayor, with republican beliefs, of his home town. Political intrigue was always a possibility, but he was well respected. If his political opponents were to remove him from office, they would have to discredit him in the eyes of the people who trusted him.

The church was powerful with the Jesuits being extremely active politically. A new, young priest came to Bourg-le-Reine and set out to destroy the mayor. Nicholas Gabriel Galois had been in the habit of making up amusing verses about some of the village characters. However, rude and vulgar verses which the cunning priest had composed were spread about the village in the name of the mayor. The elder Galois was stunned and all his political trust and confidence was destroyed. This was something which he could not bear. He came to Paris and in a room not far from Evariste's school, he took his own life.

The effect of this calamity must have been momentous. Along with the intense emotion which Evariste experienced, there was a feeling that all the principles in which his father believed had been perverted and that liberty and justice were being denied the people of France by cunning political trickery.

A bizarre final scene to the tragedy took place at his father's graveside. Nicholas Gabriel Galois was to be buried from the local church, where the offending priest was incumbent. Another priest was sought for the funeral service but the parish priest was present at the proceedings and anger arose between the dead mayor's loyal supporters and those who sided with the wily priest. This became a battle as the coffin was hastily dropped into the ground with Evariste clutching at it as it fell.

As the school year ended, Galois gained entry to the Ecole Preparetoire, a college to prepare future teachers for schools of the kind he had just left. It was at this time, little more than a senior department to his old school and it had many rules which could make large demands on the liberty of its students. Later, during Galois stay, it was renamed the Ecole Normale. Here he made a friend of Auguste Chevalier and this friendship was to last for the rest of his life. Peace and happiness were still to elude him. However, during February 1830, he prepared a new paper on his researches on the solutions of equations and submitted it for the Grand Prix de Mathematiques at the Academy of Sciences. It was accepted by Jean-Baptiste Joseph Fourier but Fourier died before he had read it. This manuscript was also lost without trace.

Galois was now very much in sympathy with the republican cause, but when the rebellion against the rule of Charles X broke out in July 1830, the director confined all the students to the school building. Students from Galois' most desired Polytechnique were on the streets fighting for the cause of political freedom, while the director of the Normale vacillated over which side he was to declare his and the school's allegiance. Finally, like all good politicians, he decided that he had been on the side of the winners all the time. The following December a letter describing the director's actions was published without name in the Gazette des Ecoles, a kind of college magazine. It was quickly attributed to Galois who did not deny the authorship and he was promptly expelled. Now there was nowhere he could develop and discipline his undoubted genius.

After his dismissal he joined an artillery battery of the National Guard and at the same time he sought the political companionship of the republican Amis du Peuple (Friends of the People). His stay in the artillery was even shorter than that at the Ecole Normale. He joined on December 4th and the artillery batteries of the National Guard were disbanded on December 31st as part of a political manoeuvre.

Galois probably had an allowance from his family, so it is doubtful whether he suffered physical hardship from the unpleasant combination of circumstances to which he had been subjected. In January 1831 he continued his mathematical activity by attempting to give a series of lectures under his own auspices in a bookshop near the Sorbonne. His intention was to lecture on his own discoveries in algebra and elliptic functions. Although these activities may have attracted a number of people out of curiosity, they do not seem to have lasted very long and it is doubtful whether they brought him much money.

Dismayed, but still determined, Galois prepared a further draft of his work for the Academy. This time Simeon-Denis Poisson read the paper and returned it with a comment explaining that he had tried "with considerable effort" to understand his work, but that it was neither sufficiently clear nor sufficiently well explained for him to judge its correctness. He indicated that it was unsuitable for publication and suggested that Galois should develop his work more and explain his theory with greater clarity.

Revolutionary politics now occupied much of this time, but his mathematical research also continued. Under the rule of Louis-Philippe there were many revolutionary incidents. People involved in such activities were arrested on numerous pretexts, police spies were everywhere and active politics was both exciting and dangerous. A particular group of nineteen prisoners was acquitted by the court and a republican dinner was given in their honour on May 9th. Galois attended this along with many others, including Alexandre Dumas, the author of *The Three*

Musketeers. During the banquet many speeches were made and much wine was drunk. As the wine took effect the speeches became less controlled. At one point Galois stood up with his glass in one hand and an open pocket knife in the other and proposed "To Louis-Philippe...". The knife wavered menacingly as the words were uttered. The gesture was obvious; the toast was not to the king's good health. Galois's toast was probably drowned by the noise and tumult of the gathering, or so the witnesses at his trial said. The next day he was arrested and imprisoned in Sainte-Pelagie.

When his trial took place on June 13th, Galois seemed concerned that he should appear a lover of the people. His attitude to the court was derisive and lacking in respect. The lawyer who defended him, and probably the jury too, showed him sympathy. The defence was simple and clever. Witnesses said that the toast was really, "To Louis-Philippe, if he turns traitor", the final words being lost to all at the banquet except those who were close to him. The knife, it was said, was for cutting his food. Galois was acquitted and left the court with all the appearances of dumb insolence and without ceremony.

Galois was at liberty but he was a marked man. He was a political agitator and thus a danger to law and order and an enemy of the crown. Whether he deliberately attempted to provoke the authorities is not certain, but his liberty was short lived. On July 14th, Bastille Day, he and a companion were leading a procession through the streets of Paris dressed in their expensive artillery uniforms. Since the artillery had been disbanded, the wearing of the uniform was illegal. The two revolutionaries were also carrying firearms.

Galois was imprisoned again in Sainte-Pelagie and it was some time before he was tried. Justice this time was harsh, his sentence was six months imprisonment. There was no allowance for the time already spent in prison, so he was not due for release until April 1832. During his stay in prison he was shot at in his cell, cruel fellow inmates got him so drunk that he was ill for days and he spent part of his time in solitary confinement in an old, damp dungeon. Galois did not serve the whole sentence in prison. A month before his release he was transferred to a nursing home at 86, rue de l'Oursine. This was supposedly to protect him from a cholera epidemic but more sinister motives could have been involved.

He was released on April 29th but was soon in trouble again. This time it was a girl. He formed a liaison with a young woman of very doubtful character. The romance cannot have lasted very long and was unsatisfactory in more than one way, because Galois describes her very harshly, and the affair landed him in a duel.

Duels could be mainly theatrical, as most were, or they could be fought in earnest. The letter Galois wrote to Auguste Chevalier the night before the event indicated that the unfortunate affair was serious and, although he had tried, he could not get out of it. There were two offended parties, and if he survived the first duel there was to be a second. Writing this letter took up almost the whole of the night before the event. He tried to detail his mathematical researches for his friend Chevalier to pass on to Gauss and Jacobi whom he believed would understand their worth. Several times he wrote, "Too little time" in the margin.

The following morning, Wednesday 30th May 1832, the duel took place in a deserted space on the outskirts of Paris. Pistols were the weapons and Galois was wounded in the stomach. Only later in the day was he found and taken to hospital. No medical attention had been given after the event and his adversaries had quit the scene.

The wound was serious. The following day Evariste Galois died after telling his younger brother Alfred that he had been murdered by police agents. Alexandre Dumas, in his memoirs, says, that the killer was called Pecheux d'Herbinville.

Evariste's funeral was not unlike his father's. Revolutionaries, police, informers and the masses attended and there was much commotion as he was buried in a common grave. A further uprising was timed to coincide with the event but this was discovered by the police and abandoned.

The important discoveries that Galois made in his research were neither read nor understood for a long time after his death. Those works published during his lifetime and those which remained unpublished but were in the hands of his friend Auguste Chevalier were published in a French mathematical journal by Joseph Liouville in 1846. From this publication mathematicians began to appreciate the worth of his discoveries.

So much so, in fact, that Camille Jordan published *Traite des Substitutions et des equations algebriques* an exposition and elaboration of Galois' ideas in 1870. From then onwards the true worth of his work was understood and the new branch of mathematics known as Galois theory began to be known and appreciated along with the mathematical structures known as Galois groups and Galois fields. These ideas were regarded as so exciting and important that the great Felix Klein, (see pages 83-97) was to remark at the end of the nineteenth century: "In France about 1800, a new star of unimaginable brightness appeared in the heavens of mathematics . . . Evariste Galois".

What were these new ideas? Why were they of such importance and why has only one mathematician been associated with the beginnings of such a momentous part of mathematics? To answer these questions we must briefly survey the knowledge of equations at the time when Galois began his studies. It has already been said that the quadratic, cubic and quintic equations could be solved by radicals, or expressions in terms of their coefficients, using only addition, subtraction, division, multiplication and the extraction of roots. The general quintic equation, containing a term x^5, was proved by Abel (and independently by Galois) to be insoluble in this way. However, there were many special quintic and even higher degree equations which were amenable to this type of solution. Galois discovered the conditions under which particular equations could be solved by radicals.

One statement, sometimes called Galois theorem, is a very precise and beautiful expression of one set of conditions for solubility. It is: If the highest power of the unknown x in an equation is a prime number (2, 3, 5. . .), and if all other values of x can be found by taking only two values of x and solving them using only addition, subtraction, multiplication and division in any combination then the equation can be solved using radicals.

There was however, a much more general method that he discovered and began to develop; this took up much of the letter which he wrote to Chevalier on the eve of his fatal duel, and was presumably described in the manuscripts lost by Cauchy and Fourier. What follows is an over-simplified form of this. First we must look at arrangements of letters or numbers known as permutations. The numbers 1, 2, 3, can be arranged 123, 132,

213, 231, 312, 321. If we call 123 the identity permutation, and keep track of what happens to the numbers by writing a top line in this order for every arrangement, then we have

$$\begin{pmatrix}123\\123\end{pmatrix}, \begin{pmatrix}123\\132\end{pmatrix}, \begin{pmatrix}123\\213\end{pmatrix}, \begin{pmatrix}123\\231\end{pmatrix}, \begin{pmatrix}123\\312\end{pmatrix}, \begin{pmatrix}123\\321\end{pmatrix}.$$

Now if we apply one permutation and then another in succession we still get one of the six

e.g. $\begin{pmatrix}123\\213\end{pmatrix} \begin{pmatrix}123\\132\end{pmatrix} = \begin{pmatrix}123\\231\end{pmatrix}$

second first

($\underline{1} \to 1 \to \underline{2}$, $\underline{2} \to 3 \to \underline{3}$, $\underline{3} \to 2 \to \underline{1}$),

this is known as closure and is the first property of a group. (Permutations also show the other group properties of associativity, identity and the possession of inverses.) Galois was one of the early pioneers of group theory and the word "groupe" is mentioned many times in his last letter to Auguste Chevalier.

If we now look at the cubic equation

$$x^3 + ax^2 + bx + c = 0 \quad (1)$$

and let the roots, or solution values of x, be α, β, γ in $(x - \alpha)(x - \beta)(x - \gamma) = 0$, then $x^3 - (\alpha + \beta + \gamma)x^2 + (\alpha\beta + \beta\gamma + \alpha\gamma)x - \alpha\beta\gamma = 0$

and comparing this with (1) we have

$a = -(\alpha + \beta + \gamma)$, $b = \alpha\beta + \beta\gamma + \alpha\gamma$ and $c = -\alpha\beta\gamma$.

Now if we interchange or permute α, β, γ, these formulae in a, b, and c will be unaltered. We can, however, make new formulae for a, b, and c, which will be altered by some permutations of the roots α, β, γ. Then the permutations which do not alter the formulae also form a group, that is, a subgroup within the original group of all possible permutations of α, β, γ and if the order of applying the permutations is unimportant that is, they are commutative as with

$$\begin{pmatrix}\alpha\beta\gamma\\\gamma\beta\alpha\end{pmatrix}\begin{pmatrix}\alpha\beta\gamma\\\alpha\beta\gamma\end{pmatrix} = \begin{pmatrix}\alpha\beta\gamma\\\alpha\beta\gamma\end{pmatrix}\begin{pmatrix}\alpha\beta\gamma\\\gamma\beta\alpha\end{pmatrix}$$

$$\left[\text{in place of } \begin{pmatrix}123\\321\end{pmatrix}\begin{pmatrix}123\\123\end{pmatrix} = \begin{pmatrix}123\\123\end{pmatrix}\begin{pmatrix}123\\321\end{pmatrix}\right]$$

and also, if these families of subgroups of permutations of α, β, γ can be made successively smaller by suitable formulae until only the identity permutation $\begin{pmatrix}\alpha\beta\gamma\\\alpha\beta\gamma\end{pmatrix}$ is left as a subgroup on its own, and if all these subgroups fit into each other in the fashion of a Chinese box, then the equation will be soluble by radicals.

Groups that have this Chinese box arrangement of their commutative, or normal, subgroups are known as soluble groups, because of their association with equations soluble by radicals. The Chinese box arrangement is known as the composition series of the group. A group can be investigated by algebraic methods to see if it is soluble, and it is known that the group of the general quintic equation, the permutation group on five letters ($\alpha\beta\delta\gamma\varepsilon$) with 120 members, is not a soluble group. Hence the general quintic equation is insoluble by radicals. Many special equations of the fifth and higher degrees have groups which are soluble and we can find a definite, but not always simple, formula for their solution in terms of their coefficients.

Galois had developed much of group theory, in particular the idea of a normal or commutative subgroup and his theory of solubility, even though he was killed a few months before his twenty-first birthday. His famous last letter to Auguste Chevalier could be claimed as the first great statement of what has become known as modern mathematics.

Sir William Rowan Hamilton (1805-1865) *BBC Hulton Picture Library*

CHAPTER TWO

Sir William Rowan Hamilton

HAMILTON WAS ALIVE at the same time as Galois, but he lived much longer and was able to realise the talent that could be seen in his youth. He was likened by some of his contemporaries to Sir Isaac Newton, the greatest of all British mathematicians. Like Newton, he made great advances in the theory of optics and mechanics but he was also very active in other parts of mathematics: he gave us the algebra of complex numbers, he contributed towards the theory of equations of the fifth degree and he added to the new field of groups. His magnum opus was the creation of quaternions. Also the concept of the 'vector' which is now very important in many areas of mathematics, was due to him.

He was born in Dublin at midnight between the 3rd and 4th of August 1805. This curious fact led him to treating the 3rd as his birthday in his youth, but in later life he acknowledged the 4th as the true date. His father was a solicitor and business man whose work kept him fully occupied and often away from home. Willy, as he was known during his childhood, soon appeared remarkable to his parents. During his second year they decided that his education was to be entrusted to his uncle and aunt at Trim, a country town about twenty miles from Dublin. His uncle was the curate of the parish and master of the local school. He was also a scholar of note and a talented linguist, with a knowledge of several languages, both ancient and modern.

Willy was a remarkable child. Only a little after his third birthday he was already a competent reader who kept his uncle and aunt occupied finding him new books. Shortly afterwards he became skilled in arithmetic and clear and accurate in geography. His greatest accomplishments, at a time when he was little more than an infant, were in languages. At four years and five months he read Latin, Greek and Hebrew. He entertained his family by reading Homer and Virgil, and he despaired at not being able to make progress in teaching Hebrew to the maid. He added the more modern languages of French and Italian to his repertoire a little later. It was only when he was about ten years old that he became accomplished in those of the Orient. His very obvious talent was encouraged by his uncle James; we may wonder what would have happened if his education had been entrusted to someone skilled in another field of knowledge. Perhaps something equally brilliant but a little more useful would have been achieved — even in nineteenth century Ireland these accomplishments must have been of limited use. For recreation he enjoyed swimming with his uncle.

When he was ten he attended school and continued his list of achievements. He wrote a grammar for the ancient oriental language of Syriac.

Although he must have been closer to his relatives at Trim than to his parents, the fact that his mother died when he was twelve must have been a saddening blow to him. Two years later his father remarried, but died himself two months later.

When William was thirteen, and again when he was fifteen, he met Zeah Colburn, the American calculating prodigy. Colburn was able to multiply together two four figure numbers correctly faster than his questioner could write them down. He could also factorise almost instantly very large numbers and perform other arithmetical tests with equal accuracy and rapidity. In a contest with Colburn, Hamilton is reputed to have fared well but to have been defeated. The two were friends and the second meeting was a happy reunion and celebration. Colburn had attended Westminster School during the intervening years. Although his arithmetical feats were prodigious, Zeah never achieved much academically. He died quite young after earning his living by a variety of jobs, finally as a rather second rate private teacher of languages.

During his school days Hamilton read of Clairaut's *Algebra* and Laplace's *Mechanique Celeste*, both in the original French. He discovered a fault in the reasoning of Laplace's verification of the parallelogram of forces. He also wrote his own textbook of algebra.

Shortly after his father's second marriage he was invited to stay with him and his new stepmother in Dublin. He cut his visit short, to only a few days, because of the pressure of his studies. The following month he returned to Dublin to present the visiting Persian ambassador with a letter (in Persian of course) written on his father's advice. His praise of the "Illustrious visitant from Iran" did not, however, get past the exalted one's secretary.

During his school days he had an interest in astronomy. He owned his own telescope and was particularly interested in eclipses. The observatory at Dunsink, which was later to be in his charge, held a fascination for him. He visited the building on at least two occasions. On the first, with an introduction from his father, he was not able to meet Dr. Brindley, the Royal Astronomer of Ireland, but was shown the instruments by his assistant. Later, on his second visit, he was warmly welcomed by Brindley himself.

In 1824, Hamilton went to Trinity College, Dublin. This gave equal weight to studies in classics and mathematics. He proved an able student. For coming first within a division of his year, a student was awarded a premium with which he could obtain books from the university bookseller. If a student came first again within the same class he was awarded a certificate. Hamilton achieved many premiums and certificates. Grades, rather than marks, were given in the examinations with the highest accolade being that of 'optime'. During his first year he achieved an optime for Greek. Later during his studies he gained a further optime in mathematical physics. To be awarded one of these honours was excellent; two was nothing short of miraculous. Soon he was highly regarded in the academic circles of Dublin.

From his childhood, Hamilton had devoted some of his time to writing poetry. His verses matured over the years but even from the beginning they showed great promise. Whether he would have been remembered as a poet if he had not achieved fame in mathematics and physics cannot be known, but he had talent and perseverance; during his first university year he was awarded two Chancellor's Prizes for his poems. Later the same year he began a friendship with the poet Maria Edgeworth.

While still at school, Hamilton had begun to study the patterns produced by rays of light on reflection, known as caustics. He extended and developed this work until in December 1824 he communicated a paper *On Caustics* to the Royal Irish Academy. The article attempted to investigate the more general properties common to all systems of light rays. A small committee of the Academy set up to referee the paper suggested that it was too abstract for publication in their Proceedings in its present form, but that it was very interesting and showed considerable analytical skill. They went on to hint that if the formulae were more fully explained and some of the conclusions were more highly developed then they would consider it worthy of publication.

Hamilton acted on this advice although he felt that the real merit of his paper had not been fully appreciated. During intervals in his studies he extended and simplified his reasoning so that a much extended work with the title *Theory of Systems of Rays* was produced and again sent to the Academy on April 23rd, 1827. The first part was published in 1828 and, even though the two remaining parts were never published in their original form, Hamilton's mathematical reputation was made. His paper was read by the astronomers and physicists of the time and hailed as a great contribution to the theory of mathematical optics.

During his student days he made many friends but there was only one romantic attachment. After a short time the young lady married someone else. Trinity College was a more intimate institution than most modern universities with students and staff well known to each other. Hamilton was well liked as a person and well respected for his intellectual gifts. During his days as an undergraduate he visited the Dunsink observatory many times and cultivated a friendship with Dr. Brindley.

Towards the end of his college course, while Hamilton was working towards the College medals for both classics and mathematics, Dr. Brindley accepted the Bishopric of Cloyne. Within the university Brindley was professor of astronomy and this position became vacant. To his surprise, Hamilton learned that if he applied for this position he would receive serious consideration. This was an extraordinary suggestion since he was still an undergraduate; and although the other applicants for the professorship included a number of Fellows of Trinity College and Airy from Cambridge, who was already a distinguished mathematician and astronomer, Hamilton received the appointment and the title of Royal Astronomer of Ireland. Hamilton was now in the strange position of being an examiner of graduate students while he was still an undergraduate himself. Not for long, however, since within a few days of his appointment he was ruled by the College as first of one hundred candidates and shortly afterwards was admitted to the degree of Bachelor of Arts.

The transition from undergraduate to professor was made easy by a long holiday. Hamilton travelled to England and his interest in poetry profited from the journey when he toured the Lake District and began a friendship with William Wordsworth. Later, Wordsworth visited Hamilton's observatory at Dunsink and persuaded him that he should devote his energies to science as he could not afford the effort it would require to be both a respected poet and a great scientist.

On taking up his appointment he moved with his sister into the house attached to the observatory. At first he embarked upon a calibration of the instruments of the observatory, but these were old and not of the highest quality and he eventually gave up this task. Hamilton did not prove to be a devoted observational astronomer and left most of this kind of work to his assistants and pupils while he devoted his time to research in optics, mechanics or mathematics. He was, however, required to give a course of lectures each year in Trinity College.

One of his pupils was Lord Adare, and when Hamilton visited the Adare country home, he was introduced to Ellen de Vere, herself of noble birth. The two struck up a friendship which developed into a romance. Hamilton was almost at the point of proposing marriage when she told him that she could not possibly leave the Curragh, where her family home was situated. This was too high a price for him to pay and the attachment ended.

In October 1832 he published a third supplement to his famous work on optics. This included the discovery of the theory of conical refraction. When a ray of light passes through water or glass it is refracted or bent. Some crystals, Iceland spar for instance, give a double refraction and two rays of light emerge instead of one. A double image can be seen if the crystal is placed over a picture or some writing. Hamilton discovered that there was no need to stop at two rays, an infinite number of rays in the shape of a cone could exist, in theory, given the right circumstances. Later his friend and colleague, the physicist Humphrey Lloyd, verified his prediction experimentally.

Towards the end of that year Helen Bayley, the daughter of a clergyman who often stayed with relatives near the observatory, began to play a part in Hamilton's life. Her ladylike appearance and religious principles appealed to him. His upbringing, with the strong Christian attitudes of his father and the direct connections with the Church and clergy through his uncle, had caused him to value religion highly. Although Miss Bayley had been known to him for some years, it was after her serious illness that summer that his feelings towards her deepened. The following year they were married, Hamilton's third romance had been a success. This success, however, turned out to be a pyrrhic victory. By all accounts the new Mrs Hamilton did not have either the physical strength nor the organising ability to run a home and family smoothly. Although conditions were tolerable in the beginning, during the later years of their marriage their homelife became very disorganised. Helen Hamilton even spent time staying with her sister to recover from her illnesses and develop enough strength and confidence to take over the activities of home and family (two sons and one daughter).

During 1833 Hamilton developed an idea of his friend John Graves on what we call complex numbers. Hamilton felt ill at ease with numbers like $a + ib$ where $i = \sqrt{-1}$ and a and b are real numbers. He could not accept that the ' + ' sign stood for addition since the real part a and the imaginary part b were entirely different things. Also he regarded $i = \sqrt{-1}$ as meaningless. His creation was to produce rules for handling couples of numbers like (a, b) in such a way that they obeyed the familiar rules of complex numbers but avoided the difficulties with i and the ' + ' sign.

If we add two complex numbers

$$(a_1 + ib_1) + (a_2 + ib_2) = (a_1 + a_2) + i(b_1 + b_2)$$

then in terms of the couples (a_1, b_1) and (a_2, b_2) the rule for addition is

$$(a_1, b_1) + (a_2, b_2) = (a_1 + a_2, b_1 + b_2)$$

A similar rule for subtraction is

$(a_1, b_1) - (a_2, b_2) = (a_1 - a_2, b_1 - b_2)$

Multiplication and division are slightly more difficult:

$(a_1, b_1) \times (a_2, b_2) = (a_1 a_2 - b_1 b_2, a_1 b_2 + a_2 b_1)$

$(a_1, b_1) \div (a_2, b_2) = \left(\dfrac{a_1 a_2 + b_1 b_2}{a_2^2 + b_2^2}, \dfrac{b_1 a_2 - b_2 a_1}{a_2^2 + b_2^2} \right)$

This algebra, which we find in present day textbooks of mathematics, has applications beyond that of complex numbers; Hamilton was aware that the geometry of the plane could be treated by algebraic couples, which we would now call row vectors in two dimensions, just as it could be done with complex numbers.

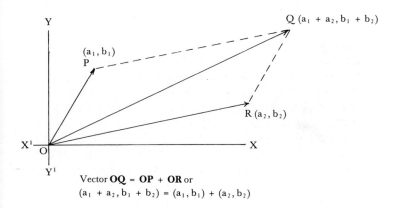

Vector **OQ** = **OP** + **OR** or
$(a_1 + a_2, b_1 + b_2) = (a_1, b_1) + (a_2, b_2)$

The obvious question which this work prompted was: Were there triples with which three dimensional geometry could be treated? The answer to this question took some time to find.

In the meantime he turned his attention to mechanics and sought to develop this mathematically in much the same way as he had tackled light rays. Some time before, Lagrange had developed a means of solving problems involving motion by considering the path of least action. Action is the product of energy and time. Hamilton had considered action in his optics and the transition to the subject of mechanics was natural. He improved upon Lagrange's least action principle because very often it did not involve least action at all. The principle which now bears Hamilton's name was a very definite improvement, and involves the action being stationary. The differential equations which need to be solved to find positions and quantities, like momentum, are much simpler in Hamilton's theory than they are in that of Lagrange. Jacobi, who was mentioned in the previous chapter, added to the work of Lagrange and Hamilton and simplifed the mathematical methods needed in this type of problem even further.

Throughout his life he was an active member of the British Association for the Advancement of Science (this rather long title is usually shortened to the 'British Association'). He attended the meetings at Oxford, Cambridge and Edinburgh in 1832, 1833 and 1834 and was deeply involved in the organisation of the Dublin meeting in 1835. At this time he was Secretary and read the Annual Report on the first evening. During the whole proceedings he acquitted himself well in his duties and had such respect as a scientist and mathematician that he was very much the centre piece of the meeting. On the last day, as a mark of his achievement and ability he was knighted by the Viceroy of Ireland, the Earl of Mulgrave, deputising for the monarch.

Two years later he received another honour. He was elected to the position of President of the Royal Irish Academy. This meant that he was regarded as Ireland's first scientist. Such recognition and reward for his work was to be seen as the zenith of a scientific career. However, the position carried with it a heavy burden of work and formal duties. One of the burdens, or joys, of such an office was attendance at dinners of his own and other scientific societies. Hamilton was one of those unfortunate people who find alcoholic drink addictive. During one meeting in Dublin he became rather the worse for drink. On the advice of a friend he abstained from potent beverages, but this only lasted for two years.

Following his work on mechanics his researches were directed towards the search for a triple or three dimensional complex number, akin to the couples which had proved useful with ordinary complex numbers. Also he examined the same problem that had occupied the time of Evariste Galois, the solution of equations of the fifth degree (containing x^5). As we have seen the mathematical community had to wait a long time for the appearance of Galois' work, so Hamilton was unfamiliar with his ideas. Abel's work, however, was fairly well known and Hamilton wrote some long papers on the subject, making some progress of his own but nothing so important as the discoveries of Galois.

Hamilton gnawed away at the notion of an algebraic triple for many years and always found that in place of developing a simple extension of the complex numbers to three dimensions he ended up with something too complicated, which was useless as a tool for three dimensional geometry. During 1843 he was hard at work on this and apparently getting nowhere, when he had one of those rare moments of inspiration. In an instant he saw something new and revolutionary, which solved his problem and was to alter the progress of mathematics.

On October 16th 1843 he was to preside at a meeting of the Royal Irish Academy. His wife had driven to meet him and the two of them were walking along the bank of the Royal Canal in Dublin. As she talked to him, his thoughts were on other things. Then, in his own words, "An electric circuit seemed to close; and a spark flashed forth . . .". In his pocket notebook he wrote down his thoughts; and as they passed under Brougham Bridge he was so overjoyed that he used his penknife to carve into the stone one line of symbols, which must be one of the most succinct and elegant in the whole of mathematics.

$$i^2 = j^2 = k^2 = ijk = -1$$

At the meeting he attended he obtained permission to read a paper at the next gathering on November 13th, on his new subject of quaternions. The stone marks on Brougham Bridge have long since been eroded but his notebook and the Council Book of the Royal Irish Academy still bear witness to the event.

What was the significance of his discovery, and how did quaternions work? Hamilton had intended to introduce a new quantity j such that $j^2 = -1$, this was intended to be analogous to i, where $i^2 = -1$. But his algebra became complicated if ij was equal to ji. We accept that $7 \times 8 = 8 \times 7$ as so obvious as to pass without comment and this commutative law of multiplication is second nature to us. So ij should equal ji by common sense; but Hamilton found that if ij = −ji instead, then his triples worked out better, but not quite well enough. Two other things had yet to come into place. His triple which was made up by a + bi + cj were a, b and c are real numbers did not appear to be sufficient to work in three dimensions. He toyed with the notion of a third quality k where $k^2 = -1$ and his electric spark of inspiration was to realise how i, j, and k were related, and that in fact a quadruple rather than a triple was needed. He gave the name quaternions to his new numbers because they were made up from four parts.

a + bi + cj + dk where a, b, c, d are real numbers.

The fundamental formula $i^2 = j^2 = k^2 = ijk = -1$ embodied the full relation between i, j and k. If we remember that the commutative rule for multiplication does not work, although a kind of anti-commutative one does, we can get the following simple formulae which we can use to perform arithmetic with quaternions. These are

ij = −ji, ik = −ki, jk = −kj
i = jk, −i = kj, j = ki, −j = ik, k = ij, −k = ji

Now if we perform some multiplications with two quaternions following these formula:

(1 + 2i + 3j + 4k)(4 + 3i + 2j + k) = −12 + 6i + 24j + 12k.
(4 + 3i + 2j + k)(1 + 2i + 3j + 4k) = −12 + 16i + 4j + 22k.

We see that quaternions themselves are not commutative. This was the price to pay for this system of numbers. However, they proved to be very useful. Hamilton spent much of the rest of his life extending them to numerous aspects of geometry and mechanics.

At the time of their discovery quaternions were a talking point of the fashionable society of Dublin. Popular articles were written about them in the Irish magazines. This was one of the rare times when a part of mathematics has been a thing to be known by informed society, even though they were not sufficiently informed to understand their meaning, significance or use. The significance for mathematicians and the development of algebra was that a sacred law could be rejected and algebra not only still work, but could be more useful. Mathematicians, with increasing bravado, experimented with new ideas and constructions which rejected the accepted laws of algebra from then onwards. In a later chapter we shall examine another algebraic system in which the commutative law does not always hold.

Hamilton felt that quaternions were a new mathematical method for geometry and physics. In fact, they were a stepping stone. Later in the nineteenth century Professor Tait, first in Belfast and then in Edinburgh championed Hamilton's cause and advocated their use in science. Others like Grassman, Heaviside and Gibbs were to develop a calculus of vectors which performed the same purpose in a way that proved more satisfactory to scientists. In a way the story doubles back on itself because Hamilton was the first to use the word 'vector'. He referred to the constant term of his quaternions as the scalar part, and that with terms in i, j and k as the vector part.

$$\underbrace{4}_{\text{scalar}} + \underbrace{3i - 7j - 6k}_{\text{vector}}$$

Also, in the preface to his first book on the subject, *Lectures in Quaternions* he gives a definition which we will recognise as similar to the one we use today: "a vector has *quantity*, in the sense that it can be doubled, tripled, etc. . . let it also be conceived to have some sense *quality* analogous to direction. . ."

Those familiar with advanced vector calculus will recognise two operations which he used:

$$\nabla = i\frac{\partial}{\partial x} + j\frac{\partial}{\partial y} + k\frac{\partial}{\partial z}; \quad \nabla^2 = \frac{\partial^2}{\partial x^2} + \frac{\partial^2}{\partial y^2} + \frac{\partial^2}{\partial z^2}$$

He believed that anyone familiar with the physics of heat, electricity, magnetism or attraction must realise the importance of these. The symbol '∇' pronounced 'nabla' was his own; he based it upon the shape of the musical instrument of the same name which is mentioned by Greek and Hebrew writers. So his classical education and facility for languages, did after all, benefit mathematics.

The famous Cayley-Hamilton theorem also originated in his work on quaternions.

In the month following the discovery of quaternions, Queen Victoria, as a mark of royal favour for his services and attainments in science, bestowed a Civil List pension of £200 per annum on Hamilton for life. Such honours were not given lightly; it was a mark of great respect to him as a mathematician and physicist. Two years later in 1845, the British Association meeting was again held in Cambridge. Hamilton attended the meeting and gave a report on quaternions, pointing out their possible applications to the theory of electric current. Later, in the summer of the same year he spent a week in the same rooms in Trinity College, Cambridge in which Sir Isaac Newton had written his famous *Principia*. Hamilton, who had been compared in stature as a scientist with the great Newton, decided to write his own great work on quaternions.

About this time and during the height of his fame, he decided to resign the presidency of the Royal Irish Academy. He felt that the time it occupied would be better spent in writing and on further research and he was also in demand to lecture on his new creation. A series of six lectures that he gave on quaternions was expanded and published in 1853 as the book *Lectures on Quaternions*. This took his ideas to a wider mathematical readership, amongst whom there were still those who doubted the wisdom of his disregard of the commutative rule. The book was not particularly successful in convincing scientists of the usefulness of his discovery, nor was it designed as a textbook. A new book, on a larger scale, was planned as a textbook to inform scientists and mathematicians of the algebra of quaternions and convince them of their usefulness as a mathematical tool. Trinity College, Dublin contributed £200 to the project, but it still absorbed much of his own finances. It had been envisaged as a work of about 400 pages, but when it was eventually published it was over 700 pages.

As the years passed Hamilton was absorbed in writing and extending his research into the applications of his creation. His wife's illness recurred frequently and the household became more and more disorganised. He would work on his own in a large room with his papers around him. Meal times were not regular; he ate when he was hungry, often disregarding the meals left for him while he was busy. During his lonely life he succumbed to the temptation of alcohol and often drank more than was good for him. He had never been afraid of work and in these later years he worked harder than ever. In his workroom papers accumulated, first on his desk and table and then on the floor. After his death, when his room was cleared, there were narrow paths between huge piles of these. Sometimes, in between the leaves were plates of food, disregarded for months, or longer.

This lonely existence contrasted with the esteem in which he was held by the academic community. When the National Academy of Sciences of the United States of America created ten foreign members, he was voted first on their list. He received honorary degrees including a Doctorate of Civil Law from the University of Durham. Several times he had been urged to allow himself to be proposed for election to the Fellowship of the Royal Society, but each time he declined. Hamilton claimed to decline this honour because of the difficulties involved in travelling to the meetings. However, his friends believed that he was not easily able to meet the expense involved.

From his early years, through the influence of his father and uncle, he was a practising Christian, a member of the protestant Church of Ireland. Even though he had had sympathies with the Oxford Movement at one time and some of his friends had been converted to the Roman Church through its influence he remained an ardent and active protestant. He was, for some time, a churchwarden.

Perhaps his last years would have been happier, or at least more comfortable and better regulated, if his wife had been able to see to the organisation of the household and create some kind of daily routine with the creature comforts that would have made his task a little easier. However, Hamilton was naive in some ways and perhaps brought difficulties upon himself. A story of his gullibility concerns the cow he kept at one time on the meadow adjoining the observatory. A farmer persuaded him that his cow would be happier with company and the same farmer agreed, on the payment of a small sum by Hamilton, to lend him his herd to graze on the same meadow as company for his cow.

In 1865 he was the victim of a severe attack of gout, a consequence of his drinking, which lasted for some months until he died on 2nd September 1865.

His *Elements of Quaternions* was published shortly afterwards. The 500 copies originally printed soon became very rare. However, it has been reprinted several times and is still available today.

Quaternions was not the only subject of his later work. His *Icosian Calculus* which absorbed part of his time is the starting point for what are now known as Hamiltonian groups and Hamiltonian graphs.

Much of Hamilton's work and his 200 notebooks are preserved and can still be seen at Trinity College, Dublin.

CHAPTER THREE

George Boole

IN HIS RESEARCHES the mathematician looks for beauty as well as for the possible applications of his discoveries. The engineer or scientist looks to mathematics to provide him with the methods by which he can solve his problems. When these qualities come together in a particular mathematical idea something very wonderful has been achieved. One person who achieved this unique combination not once, but twice, was George Boole.

All branches of engineering and science abound with problems that are most conveniently formulated in terms of differential calculus. Many of these problems are eventually expressed as differential equations. A particular convenient method for a solution of certain types of these was one of Boole's major discoveries and it won him a gold medal from the Royal Society. His own development of this idea is still used by sixth form and university students.

During this century electrical equipment has developed to the extent that a design may involve thousands of circuits, each with numerous components. To develop the simplest and least expensive circuit to perform a particular function is very important. So a mathematical means of describing a circuit is of great practical importance to a design engineer, especially if the mathematics can be simplified by a small number of easily understood rules to show the simplest equivalent circuit. Boolean algebra, a direct descendent of the logic of George Boole fulfils this purpose.

Boole's logic was, however, much more than a mathematical

George Boole (1815-1864) *Mansell Collection*

method that was destined to become an important tool for the electronic engineer, it was a significant step in the development of what is now known as symbolic logic. Except for the work of Leibniz, logic had made little real progress from Aristotle until the discoveries of Boole and De Morgan, with Boole making by far the greatest contribution.

George Boole was born at Lincoln on the 2nd November, 1815. His father worked as a cobbler but was very interested in scientific pursuits and generally valued all knowledge. With him, George made various scientific devices. They were particularly adept at making optical instruments and their telescope was a great success. In his shop window George's father put a notice inviting anyone to look at the works of God through his telescope. He was also closely involved in the formation of a Mechanics Institute in the city. Mechanics Institutes were associations where people with a little formal education could gather together in the evenings to learn about the scientific principles (and perhaps more basic knowledge) on which their own trades were based. Although their aim seems modest now, they had a considerable effect on the education of the working people and on the spread of knowledge. Many trades, through the increased skill of their craftsmen, benefited from these organisations. Boole's father was also interested in mathematics, religion and philosophy and passed on his enthusiasm for these to his son.

George went first to a National School. Schools of this kind no longer exist, but their style of teaching is still worth describing. One master trained his senior pupils as monitors who in turn drilled the younger and less able children in the basic skills of reading, writing and arithmetic. The monitors received lessons for a small part of the time from more senior monitors, junior teachers or perhaps from the master himself. One proponent of this system of education boasted that 100 children could be taught by one teacher. Rote learning was the stock in trade of these schools and little real education suitable for a person with a desire for knowledge and a facility for thinking can have taken place.

On leaving the school Boole spent a short time at an institution specialising in commercial subjects. When he was freed from the stifling effect of mechanical drilling his desire for knowledge intensified. With the help of a local bookseller, who later became a close friend, he learned Latin. On mastering this he went on, not unlike Hamilton, to learn Greek, French and Italian. When he was fourteen his father sent his translation of the Greek poet Meleager's *Ode to Spring* for publication in a local newspaper. This was Boole's first published work. However, criticism came his father's way when a schoolmaster who read the translation accused him of misrepresentation since he did not believe that such a fine piece of work could be that of a mere youth, especially one without the benefits of a grammar school education (where Latin and Greek took up a large part of the curriculum).

In July 1831, when he was sixteen, Boole obtained the position of usher in Mr. Heighing's School at Doncaster. An usher was the second or junior teacher in a school and often his duties included many of those now undertaken by a caretaker. This was a boarding school and the number of boarders rose from 12 to 55 during the two years of Boole's stay, so he must have had very little time for himself. But it was in his spare minutes that he began his study of mathematics. He claimed to have taken up the subject because he did not have access to a good library and mathematics textbooks were the cheapest source of knowledge since they took longer to study than any others. Over the next few years he studied the books of the French masters Lacroix, Poisson and Lagrange. One of his students at this school said that he was a very fine teacher with able students, but very impatient with the rest. His great desire, it was said, was to find a few minutes, particularly when his charges were writing, to study his own mathematics book.

Most schools at this time were based on religious principles. This school was Methodist (Wesleyan) but, even though Boole had intended to prepare himself for ordination as a clergyman, he was at odds with the Faith of the school. He leaned towards the Unitarian beliefs that God, Jesus and the Holy Spirit were not one and the same and that the Bible was not necessarily the whole truth. It was probably because of this that he left and went to another school in the village of Waddington, before opening a school of his own in Lincoln when he was twenty years old.

His knowledge of mathematics was respected by the people of Lincoln through his work as a teacher and because of his connections with associations like the Mechanics Institute so much that when he was nineteen he was asked to write an address on Newton for the occasion of the presentation of a bust of the great man to a local Institute. This address was printed in a local paper. Boole's study of mathematics progressed but he was later to complain that he had lost five years of time because he taught himself. He believed that had he had the right kind of guidance he could have made better progress. However, during his early twenties he made discoveries and formulated ideas that he was to publish later.

At this time there were very few scientific journals in which to publish research. The senior learned bodies, especially the Royal Society, published scholarly work, but usually the submission of research needed to be sponsored. The quality of work selected was very high and these were not places for the unestablished mathematician to begin. In 1837 the Cambridge Mathematical Journal was founded and Boole published his first learned paper in this in 1840. The editor was Duncan F. Gregory, himself a mathematician of note and grandson of James Gregory the discoverer of the series of that name. Gregory appreciated Boole's work and encouraged him and helped him to prepare for publication. His early published work included the idea of a mathematical invariant, a notion which had escaped even the greatest of the mathematicians of the time but was later to be developed by Cayley, Sylvester and Hilbert.

Cambridge was obviously the place for an aspiring mathematician, but Boole was without a degree or even formal training in the subject. Even with the eventual hope of winning a fellowship and following the scholarly life as an academic research mathematician, he would first have to become an undergraduate and take a degree. Friends encouraged him in this course, but he consulted Gregory who suggested that he would have to give up the development of his own ideas and take up a very disciplined study of the subject. Gregory suggested that this might stunt his mathematical development rather than enchance it, since success in the course and examinations at that time were as much dependent on learning well known difficult methods as on real mathematical ability. Another factor which weighed heavily against him going to Cambridge, was the expense involved and that his elderly parents were now at least partly dependent on him. Eventually he decided not to go and fortunately he never had cause to regret this decision.

Just before his death in 1844, Gregory recommended that Boole should submit his paper *A General Method of Analysis* to the Royal Society. Gregory himself guaranteed it his sponsorship if Boole had difficulty in getting his work accepted for publication. In fact it was published in the *Philosophical Transactions* that same year. This was to be the making of Boole; his new and original method was there to be read and admired by the world's mathematicians, and it was the cause of the Society's award of their gold medal for the most important paper, on any subject, published during the previous three years.

Before we carry on with the sequence of events which was to aid his career to the zenith of the academic profession we will examine the discoveries that first made his reputation.

In the differential calculus it is common to write $\frac{dy}{dx} = f'(x)$ when differentiating $y = f(x)$. We know that $\frac{d}{dx}$ operates on y with the process of differentiation. Boole separated $\frac{d}{dx}$ from the expression and treated it on its own. He developed an algebra in which expressions in $\frac{d}{dx}$ can be handled and manipulated, according to certain rules, to a form more amenable to the solution of problems.

Although Boole used D to represent something slightly different, it is now quite common to use D as the differential operator $\frac{d}{dx}$ and D^2 as $\frac{d^2}{dx^2}$. These are commonly used in the solution of linear differential equations by the method developed by Boole.

An equation $\frac{d^2y}{dx^2} - (\alpha + \beta) \frac{dy}{dx} + \alpha\beta\gamma = f(x)$, ($\alpha$, β are constants) can be expressed as $(D^2 - (\alpha + \beta)D + \alpha\beta)y = f(x)$.

Now, $D^2 - (\alpha + \beta)D + \alpha\beta = (D - \alpha)(D - \beta) = F(D)$ say, so $(D - \alpha)(D - \beta)y = f(x)$. It is known that Ae^α and Be^β (where A and B are constants) are solutions which can be found by solving $F(D) = 0$ as a quadratic equation. The story does not end here, since these two solutions only make up the part of the complete solution known as the complementary function. It is also necessary to find a particular integral of the equation. Boole gave a general method of inverting the operator expression and a number of particular theorems for different types of $f(x)$. One such theorem is $F(D)e^{ax} = e^{ax} F(a)$, here $f(x) = e^{ax}$.

As an example consider the method applied to the equation

$$\frac{d^2y}{dx^2} - 3\frac{dy}{dx} + 2y = \frac{1}{5}e^{3x}$$

Rewriting we get $(D^2 - 3D + 2)y = \frac{1}{5}e^{3x}$

$$(D - 2)(D - 1)y = \frac{1}{5}e^{3x}$$

Hence $\alpha = 2$, $\beta = 1$ and the complementary function is $Ae^{2x} + Be^x$

For the particular integral:

inverting, $y = (D^2 - 3D + 2)^{-1} \frac{1}{5}e^{3x}$

$$y = \frac{1}{D^2 - 3D + 2} \frac{1}{5}e^{3x}$$

Applying the theorem $F(D)e^{ax} = e^{ax} F(a)$ gives a particular integral of the form

$$\frac{1}{a^2 - 3a + 2} \frac{1}{5} e^{ax}$$

$$= \frac{1}{3^2 - 3.3 + 2} \cdot \frac{1e^{3x}}{5} \quad \text{(where } a = 3\text{)}$$

$$= \frac{1}{10} e^{3x}.$$

Hence the complete solution is $y = Ae^{2x} + Be^x + \frac{1}{10} e^{3x}$.

For some years Boole had been interested in logic. His aim was to develop a philosophical logic, which allowed thought to be expressed and manipulated by mathematical means. His ideas were developed over quite a lengthy period but certain key points came together in a flash of inspiration. Even so, he later claimed that his first work on the subject, a pamphlet entitled *The Mathematical Analysis of Logic* was published before he had fully explored the consequences of his system. The event that precipitated his setting down of his ideas for publication in 1847 did not really concern him directly, but so pressured him into action that he commented on it in the introduction to his first short book.

Boole was in the habit of reading philosophical journals, and in the spring of 1847 an academic argument over a point of logic was being carried on by published letters between Augustus De Morgan and Sir William Hamilton. De Morgan was a highly respected English professor of mathematics who was attempting to regenerate interest in logic from the point of view of mathematics. This Hamilton was not the Sir William Rowan Hamilton of Dublin, but a Scottish philosopher of high standing who was approaching the subject from a different point of view. The actual substance of the argument is of no importance now but what is still remembered is the Scot's disregard and belittlement of things mathematical. Two quotations of Hamilton are: ". . . in mathematics dullness is thus elevated into talent, and latent disregard into incapacity . . ." and ". . . mathematics cannot conduce to logical habits at all . . .". Boole sums up Hamilton's views by writing "Sir W. Hamilton has contended, not simply, that the superiority rests with the study of logic, but that the study of Mathematics is at once dangerous and useless". This is a point of view, but one which few people would agree with now, and I doubt very much if he found many real supporters then. Certainly De Morgan, who was a man of both genius and wit, put him very definitely in his place.

This argument awakened interest and set the scene for a development in logic to be recognised and appreciated. Boole realised that logic could be viewed from the notion of quality and could be based on a system of relations which belong to the mind. "The mathematics we have to construct are the mathematics of the human intellect".

In this system the letters x, y, z are sets, Boole called them classes. 1 is the universal set and 0 is the empty set. Now, for example, xy is the set of grey haired men if x is the set of grey haired things and y is the set of men. Plainly xy = yx, and this logical multiplication is commutative. We can easily extend this idea to the set xyz, grey haired men over thirty, if z is the set of things over thirty. The set x + y is the set of things which are x or are y. The set of grey haired things or men. Again the commutative law holds, x + y = y + x. The distributive law of multiplication over addition also holds since x(y + z), the set of grey haired things that are also men or over thirty, is the same xy + xz, as the set of grey haired men or grey haired things over thirty. That is x(y + z) = xy + xz.

The one important rule which we do not find among the laws of ordinary algebra is x^2 = x. If written x.x = x, it is obvious that the set of grey haired things and the set of grey haired things is still the set of grey haired things. Boole even extended this to x^n = x where n may be greater than 2. From x^2 = x we get x^2 − x = 0 or x(1 − x) = 0. Since 1 is the universal set, 1 − x is the set of those things which are not grey haired. x(1 − x) is therefore the set of those things which are grey haired and not grey haired. Clearly this is the empty set, designated as 0. Extending our example we can have y(1 − x), the set of men without grey hair, and y(1 − x) (1 − z) the set of men without grey hair who are thirty or under. Subtraction, as exemplified by x − y, means the set of grey haired things which are not men. Note that y − x is the set of men who are not grey haired, and that (1 − z) (y − x) is the set of men thirty or under who are not grey haired. These ideas could also be used to simplify expressions or obtain alternative descriptions.

Boole's system did not change very much from this, even though it was explained much more fully and applied to other areas, particularly the study of probability, in his later writings. Modern Boolean algebra is an extension of Boole's system and will be discussed later.

The Mathematical Analysis of Logic was sent to the Rev. Charles Graves, Professor of Mathematics at Trinity College, Dublin for his suggestions and general approval, which was forthcoming. Whether or not the book sold well, it was remembered in the right places. It was Boole's second major development in mathematics and was a great recommendation for him in being chosen as Professor of Mathematics for the newly founded Queen's College in Cork. The self-taught, self-made mathematician, through his own ingenious endeavours, made it to the top of the academic tree after eighteen years as a teacher in small, rather insignificant schools.

Up to the time of his departure for Cork in 1849 Boole helped the working people of Lincoln in many practical ways. He taught classics and mathematics at the Mechanics Institute, and he was a trustee of an institution for female prisoners. The Early Closing Association of the same town was an organisation with the aim of reducing the hours of work of the working people. Boole was vice president of this and the author of *The Right Use of Leisure*, which was first given as a printed address and was later reprinted by the Association for the benefit of the townspeople. Just before he left for Ireland, a public supper over which the mayor presided, was given in his honour, for his services to the city.

Once in Cork, he had a very different life. Freed from the drudgery of teaching elementary subjects, he had the time and the more academic atmosphere in which to pursue his researches into mathematics, his writing and his reading of philosophy. He extended his earlier pamphlet on logic into an extensive treatise involving philosophical considerations as well as mathematics. This new work *An Investigation of the Laws of Thought on which are founded the Mathematical Theories of Logic and Probabilities* was published in 1854, mainly at his own expense. The real advance he made in this was the extension of his logical algebra to the study of probability, but earlier ideas were also more fully explained. This book, like the earlier one, is now a classic, but whether at the time he made a profit financially is doubtful. There was no doubt, however, that his work was recognised as he was honoured by the universities of Dublin, Edinburgh and Oxford during the 1850's and in 1857 he was elected a Fellow of the Royal Society.

In 1859 his book *Differential Equations* was published, and it is a very fine work on the subject. It contains a full exposition of his differential operator algebra and the process of inversion in the solution of linear differential equations. Only with the appearance of A. R. Forsyth's books on differential equations at the turn of the century, was Boole's work displaced from being the standard textbook on the subject. The following year his *Finite Differences*, another textbook that has enjoyed substantial popularity, was published.

In 1855 Boole married Mary Everest, the niece of the man after whom the mountain was named. Her father was an eccentric clergyman from Gloucestershire and another uncle of hers was Professor of Greek at Cork.

Boole had an affinity with classical languages, and held Professor Ryall, her uncle in Cork, in high esteem; indeed, *The Laws of Thought* is dedicated to him. There were five daughters to the marriage; Alice was gifted mathematically, and constructed certain models well ahead of their rediscovery by the Dutch mathematician Schrute; and Lucy became the first woman professor of chemistry in England. Apart from bearing these children, Mary Boole was a doting wife who never ceased to sing the praises of her husband, before and after his death. Boole was concerned that the minds of his daughters should not be corrupted and although he was a very religious man, he shielded them from the influence of the churches and clergymen because he wished them to make up their own minds free from their persuasion.

His faith as a Christian was perhaps intellectual rather than emotional. Earlier it was mentioned that he was of Unitarian persuasion, but during these later years in Cork he was reluctant to discuss the subject; a wise decision in view of the religious differences in Ireland.

The people of the town of Cork regarded him as half a saint and half a fool. He was considered too simple and naïve to cheat. Those of all social strata and religions were befriended by him.

All were welcomed in his home. Like his father, he had a telescope and allowed anyone who wished to view the heavens through it. Many of the people who knew him, such as neighbours or shopkeepers, never suspected that he was a man of genius and fame, and some believed that because of his naïve trust he must be a simpleton. As he had done for the people of Lincoln, so he attempted to do for the people of Cork by helping with organisations for the welfare of the working people.

As a university teacher he was well respected; once more he showed his ability to develop the minds of able students. He was reputed to construct his lectures as if he was discovering the subject for the first time himself. He did not appear to lecture his students on mathematical truths, rather he placed them as participants in an act of discovery.

His high regard for beauty was shown, like that of Hamilton, by the writing of poetry. Boole's poems were not as good as those of the other Irish mathematician, but they gave him much pleasure in their writing; until his wife said that she thought he should concentrate on his great discoveries in mathematics. In this he also regarded beauty as important: "However correct a mathematical theorem may appear to be, you ought never to be satisfied there is not something imperfect about it till it gives you the impression of being also beautiful".

Towards the end of 1864, Boole, after being soaked and chilled by the winter weather, caught a cold which eventually turned to pneumonia. Typical of Boole's lifelong support for the underdog, the doctor he called was a man who had recently been dismissed from his position as professor of medicine at Queen's College for some kind of misconduct. He did this to show support for a friend who was now deprived of his livelihood. Unfortunately, his condition worsened and he died on 8th December 1864.

Boole was very much the self-made mathematician. With no formal training in the subject he reached the pinnacle of academic success with his appointment as professor and his election to the Royal Society. After his death, the Cathedral at Lincoln installed a stained glass window in his honour. Queen's College similarly honoured him, with a window in College Hall.

His widow devoted her life to his memory and her own somewhat misguided extensions of his ideas. She gathered about her a number of disciples who listened to her interpretation and application of her husband's discoveries and she published such books as *Logic Taught by Love* and *The Philosophy and Fun of Algebra*. Mary Boole could not do any real harm to the appreciation of George Boole's work, it was too great for that. His logic was such a fruitful area for research and application in mathematics, and later engineering, that very soon it was developed and extended beyond its original conception.

The obscurities and difficulties which exist in Boole's work were spotted and clarified by a number of logicians, who went on to construct an extensive logical system. Among the early extenders of Boole's ideas were W. S. Jevons, E. Schroder and the American logician, C. S. Pierce. Schroder and Pierce were responsible for the very peculiar distributive law of addition over multiplication, which holds true in Boolen algebra and certain other logics but does not work in our ordinary algebra.

$$x + y.z = (x + y)(x + z).$$

Our modern system of Boolean algebra really stems from the work of E. V. Huntington who first in 1904 and then nearly thirty years later in 1933 published no fewer than six alternative expressions of the axioms and theorems of this mathematical structure. Modern summaries of the laws and rules of Boolean algebra are usually based on these. Such a summary is:

Commutative Laws $\quad x + y = y + x, \quad x.y = y.x$

Associative Laws $\quad x + y + z = (x + y) + z$
$\quad\quad\quad\quad\quad\quad\quad\, x.y.z = (x.y).z.$

Distributive Laws $\quad x.(y + z) = x.y + x.z$
$\quad\quad\quad\quad\quad\quad\quad\, x + (y.z) = (x + y).(x + z).$

Sum Rules $\quad x + x' = 1$ (where x' is the complement of x,
$\quad\quad\quad\quad\quad\quad\quad\quad\quad\quad$ or not x')
$\quad\quad\quad\, x + 1 = 1, \quad x + 0 = x, \quad x + x = x.$

Product Rules $\quad x.x' = 0, \quad x.0 = 0, \quad x.1 = x,$
$\quad\quad\quad\quad\quad\, x^2 = x.x = x.$

Absorbtion Rules $\quad x + x.y = x, \quad x + x'.y = x + y.$

De Morgan Laws $\quad (x + y)' = x'.y', \quad (x.y)' = x' + y'.$

Sometimes, especially when the algebra is used to manipulate sets or logical propositions the signs ∩ or ∧ replace . and ∪ or ∨ replace + . Although Boole's system was originally conceived as applying to classes, which we would usually interpret as sets, modern uses of the algebra include applications to electrical circuitry where x stands for a switch in the 'on' position and x' for one in the 'off' position. Also '0' can represent 'no current' in a circuit and '1' the presence of 'current'.

One modern application is to use the algebra as a calculus to simplify the representation of a circuit to one of the simplest equivalent circuits, which will be easier and cheaper to construct, and ultimately more reliable.

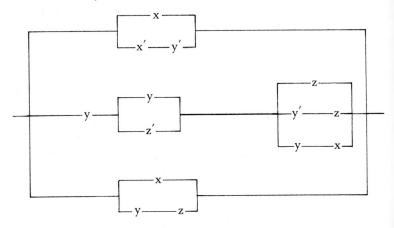

We can represent the circuit C in Boolean algebra by

$$C \equiv x + (x'.y') + y.(y + z').(z + y'.z + y.x) + x + y.z$$

$$C \equiv x + x'y' + y^2z + yz'z + y^2y'z + yy'zz' + y^3x + y^2z'x + x + yz$$

$$C \equiv x + x'y' + yz + yx + yz'x$$

$$C \equiv x + yz + x'y', \text{ which cannot be further reduced.}$$

Hence

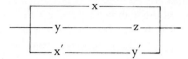

is equivalent to C.

The operations which electronic digital computers perform depend on the manipulation of electrical circuits by gates. The behaviour of these can be described in terms of Boolean algebra with '1' indicating 'current' and '0' 'no current'. Hence the algebra can play a major part in the design of the logical circuits for particular purposes. For instance, addition circuits are constructed from gates with their precise requirements calculated by Boolean algebra.

Most circuits are constructed from a number of these logic elements. The description of a number of different types of gates are given below with tables indicating their behaviour under various current inputs. These tables are directly analogous to the truth tables used in evaluating logical expressions.

And   *Or*

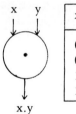

x	y	x.y
0	0	0
0	1	0
1	0	0
1	1	1

x	y	x + y
0	0	0
0	1	1
1	0	1
1	1	1

Not

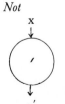

x	x'
0	1
1	0

Gates performing these functions can be connected together to produce the arithmetic circuits of a computer. However, greater simplicity and flexibility can be made in the design of a computer by using gates which are all of the same logical type.

A circuit for any specific function can be obtained using combinations of only one type of element: "or and not" or NOR, or "and and not" or NAND.

NOR *NAND*

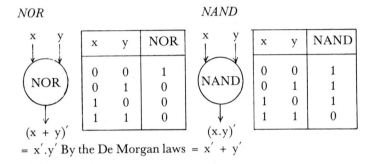

= $x'.y'$ By the De Morgan laws = $x' + y'$

Well over 100 years ago George Boole believed he had constructed the mathematical description of the laws of thought. If this is now viewed as something of an ambitious assumption for the mind of man, then certainly he made a great step in the description of the laws of 'thought' of the electronic digital computer.

CHAPTER FOUR

Arthur Cayley

DEVELOPMENT IN MATHEMATICS does not always come about because of a proof of a theorem, or the insight to see a connection between two unrelated areas that benefits them both; sometimes it comes about in a much simpler way. When a page of equations can be condensed into a few simple symbols, or even only one, which can be used and manipulated to give the same effect, then something very useful has been added to the mathematician's armoury. Not only is paper, time and work saved but, much more importantly, tedious, non-creative, but necessarily accurate brainwork is considerably reduced.

Arthur Cayley was a person whose output of mathematical writing was one of the largest of all mathematicians. There was scarcely an area of mathematics or mechanics to which he did not make a significant contribution. However, the most widely known creation of his is purely a shorthand way of writing that which was already known. He invented matrices, developed the rules for their manipulation and discovered how they behaved as an algebraic structure.

Cayley's father was engaged in trade in Russia and the family lived in St. Petersburgh, returning to England periodically. During one of these visits Arthur was born in Richmond, Surrey on 16th August 1821. The family went back to Russia for eight years and then returned to England in 1829 when Arthur's father retired from business. They settled in Blackheath, where Arthur went to a private school before going on to King's College School, London at the age of fourteen. During his early school days he showed his natural ability for mathematics. He took

Arthur Cayley (1821-1895) *Trinity College, Cambridge*

ARTHUR CAYLEY 47

great delight in arithmetic and developed great speed and skill in long numerical calculations. At King's College School he proved to be an able student in most subjects, but particularly in mathematics and sciences. His father wished him to enter his own former business and engage in foreign trade, which would mean the end to his mathematical pursuits. Eventually the Principal of his school persuaded Arthur's father to allow him to go to Cambridge and realise his mathematical potential.

In October 1838 he entered Trinity College, Cambridge at the age of seventeen. Cayley's ability was soon noticed and he progressed well, winning a college scholarship in 1840 at the earliest time open to him. One year, in his annual examination, he scored more than twice the marks of the next most able candidate. In 1842 he graduated as Senior Wrangler, those who obtained honours degrees in mathematics were known as Wranglers; the first, with the most marks, was known as the Senior Wrangler and the rest were numbered in order of merit. Later he won the first Smith's Prize for mathematics.

As an undergraduate Cayley had been renowned for his hobby of novel reading. This persisted throughout his life and he gained much pleasure from the novels of Sir Walter Scott and Jane Austin. He disliked those of Charles Dickens intensely. Reading was not his only hobby, he enjoyed painting, mainly with water colours. During his youth and early manhood he was an active traveller, walker and mountaineer. He spent holidays in the mountainous parts of Europe, especially Switzerland and Italy. On these visits to Italy he also spent time admiring the paintings and architecture. One of London's famous gentlemen's clubs, the Alpine Club was founded in 1857 for those interested and active in mountain climbing. Cayley was one of its earliest members.

He was elected to a minor Fellowship of Trinity College in 1842. In this position he had few obligations to either the College or the university and was free to coach students and pursue his own mathematical researches. It was usual for mathematics students to be coached in the solution of problems by private tutors since the university provided only theoretical lectures, but the examinations demanded skill in the art of problem solving. Cayley took pupils, and had a considerable mathematical output of his own. During this period he visited Dublin and heard

Hamilton's lectures on quaternions.

His published research began when he was still an undergraduate, with three articles in the newly established 'Cambridge Mathematical Journal'. During his early years as a Fellow he published twenty-five papers. In 1845 he was elected a major Fellow of his college, but the joys of this position could not be enjoyed indefinitely. He could not retain his Fellowship for more than seven years after taking his M.A. without taking holy orders. Although he was a member of the Church of England he had no desire to be ordained; his respect for the Church was sufficient to prevent him taking orders only as a means of personal advancement.

The limit of his period of Fellowship was 1852, but in 1846 he left Cambridge and took the law as his profession. He entered Lincoln's Inn and became a pupil of Mr. Christie, who specialised in conveyancing, the legal work entailed in transferring the ownership of property from one person to another. As a pupil he was able and diligently attended to his work. He enjoyed the company of his fellow pupils and did not let his honours at Cambridge separate him from them. At the same time he did not relinquish his work on his real love, mathematics. On 3rd May 1849 he was called to the bar. He showed such skill in his work that he was never short of clients, most of whom he turned away and some of his drafts have been quoted in law textbooks as model examples. He accepted work only through Christie and then only sufficient to give him an income equal to his needs. Rather than attempt to make a fortune, he preserved as much spare time as he could for his research.

During the fourteen years he spent at the bar he produced well over two hundred mathematical papers, very often opening up whole new fields of knowledge. Near Lincoln's Inn lived his friend James Joseph Sylvester, who had had a chequered career as a mathematician. First, because he was a Jew, he was unable to be admitted to a degree at Cambridge, and later he experienced so much difficulty in obtaining suitable academic positions that he became employed as an actuary by an insurance company. The two spent much time together, oblivious to legal matters and day-to-day problems, discussing their ideas and discoveries.

If Cayley chose not to acquire a high legal reputation, he

certainly did not lose his respect as a mathematician. In 1852 he was elected a Fellow of the Royal Society because of his achievements. Also, on a number of occasions, he was called back to Cambridge to act as an examiner.

Cayley's salvation from the legal profession came about in 1863 when he was elected as the first Sadleirian professor at Cambridge, with the duties "to explain and teach the principles of pure mathematics and apply himself to the advancement of science". This newly created position came about from the bequest of Lady Mary Sadleir. Over 150 years earlier money had been left to the university to endow lectureships in algebra in nine of the colleges. Since their establishment, mathematics had changed to such an extent that these positions were inadequate for the development of mathematics as a whole, and the lectures had generally ceased to attract students. To Cayley's benefit, the university reorganised the statutes concerning the endowment and created the Sadleirian professorship, which still exists today.

He had no hesitation in accepting this appointment, which allowed him to devote himself completely to mathematical research and teaching, even though the salary was very modest and much less than he was earning by practising law. Later, the rules were again altered and he received a better income for an increase in his lecturing duties.

Shortly after returning to Cambridge, where he remained for the rest of his life, he married Susan Maline and settled into a peaceful domestic and academic life.

His lecture courses were usually very advanced and attracted few undergraduate students. Those who did attend were amongst the most gifted who passed through the university at that time. Mostly he lectured to small audiences of advanced students and established mathematicians. Since his students were few in number, he usually dispensed with the blackboard and brought most of his lecture material already prepared on the back of paper that was used for making blueprints. (This was the usual kind of paper used at that time for working out problems by mathematics students.) His diagrams were beautifully prepared, showing off his skill in water colour painting.

Cayley's output of research papers was still very large. He often wrote in French and published his work in France. His only book, *A Treatise on Elliptic Functions*, was published in 1876. This was intended to make elliptic functions accessible to English mathematicians, since most of the works on the subject were written in either French or German.

Over the years, his skill as an administrator and his knowledge of business and the law was sought on many occasions. Undoubtedly this was a more pleasant use of the skills developed in Lincoln's Inn than he had previously experienced, but it was still his ability at drafting documents which caused colleges, the university and scientific societies, to seek his advice.

He was a man of gentle disposition, but he was prepared to work tenaciously for the causes he believed in. As a quiet champion of higher education for women, he devoted time to the ladies colleges in the university. In Girton College he lectured on mathematics, and he was chairman of the Council of Newnham College.

During 1881 he was invited to give a course of lectures at Johns Hopkins University, Baltimore where his friend Sylvester was now professor of mathematics. This invitation was a high honour and Cayley spent five months in the United States of America. The substance of these lectures, like that of many other courses of lectures, appeared as a lengthy article, published in the scientific press. Nearly all his courses of lectures were based on the research he was pursuing at the time. It was usually the delight of his audiences to find that a lecture was based on discoveries he had made since the previous one.

Like Hamilton and Boole, he was a member of the British Association, but unlike Hamilton he was elected to the Presidency. In 1883 he gave his Presidential address in Southport. It was the custom of most presidents to speak on a rather general topic, which was at a level where the subject matter could be appreciated by most of the audience, which included reporters, laymen and specialists in other fields. Cayley dispensed with this tradition and gave a talk that was strongly mathematical in content but contained some elegant passages of a more generally understandable nature. The press, especially the leader writers, praised it warmly and widely quoted his description of modern mathematics:—

"It is difficult to give an idea of the vast extent of modern mathematics. The word 'extent' is not the right one. I mean extent crowded with beautiful detail — not an extent of mere uniformity such as an objectless plane, but of a tract of beautiful country seen at first in the distance, but which will bear to be rambled through and structured in every detail of hillside and valley, stream, rock, wood and flower. But, as for everything else, so for the mathematical theory — beauty can be perceived but not explained."

Throughout his life his respect in the mathematical community and in academic life generally had been held in high esteem. He was awarded medals by the Royal Society and honorary degrees by many British and foreign Universities, and he was an honoured member of many scientific societies throughout the world.

Even in his later years, his interest in painting, the work of the master artists and architecture grew rather than diminished. His knowledge of languages was considerable. He read French, German and Italian and wrote in French as well as he did in English. He knew Latin and Greek and read the Greek plays in their original language.

As well as high level research he took a keen interest in book keeping, no doubt a legacy from his days as a pupil in Lincoln's Inn, and in his British Association address he compared double entry book keeping with the use of positive and negative numbers. Book keeping was regarded by him as one of the perfect sciences, and he wrote a pamphlet on it some months before he died.

In January 1895 he suffered the last of a series of severe illnesses and died on the 26th of that month at the age of 73.

The 13 volumes of his collected works contain 967 papers. There was scarcely a branch of mathematics that he did not influence, but he is particularly remembered for his work on algebraic invariants, these are expressions which allow the shape of a geometrical figure or the form of an algebraic equation to remain unaltered by a mathematical transformation. This field was originated by Boole, and Cayley gained support and encouragement from Sylvester, whose own researches were in this area. The geometry of n-dimensions was pioneered by Cayley, but the methods now used in the subject are different from those he developed. His invention and description of matrices, however, has remained almost unchanged since he first published the notation in a paper written in French in 1855. His fully developed ideas appeared in *A Memoir on the theory of Matrices*, published 3 years later in 1858.

Matrices are of considerable importance to schools and colleges and in the application of mathematics in science, industry and commerce. They have applications from the design of aeroplanes, to how a travelling salesman should plan his route.

Cayley showed that a matrix such as $\begin{pmatrix} 2 & 3 \\ 5 & -1 \end{pmatrix}$ would arise naturally from the consideration of the simultaneous linear equations:

I = 2x + 3y
II = 5x − y

which can also be represented as

$$\begin{pmatrix} I \\ II \end{pmatrix} = \begin{pmatrix} 2 & 3 \\ 5 & -1 \end{pmatrix} \begin{pmatrix} x \\ y \end{pmatrix},$$

with only a slight change to Cayley's original notation.

Now $\begin{pmatrix} 2 & 3 \\ 5 & -1 \end{pmatrix}$ can be removed from this system and handled separately.

Another matrix with the same number of rows can be added to it or subtracted from it.

e.g. $\begin{pmatrix} 2 & 3 \\ 5 & -1 \end{pmatrix} + \begin{pmatrix} 1 & 2 \\ -2 & 3 \end{pmatrix} = \begin{pmatrix} 3 & 5 \\ 3 & 2 \end{pmatrix}; \quad \begin{pmatrix} 2 & 3 \\ 5 & -1 \end{pmatrix} - \begin{pmatrix} 1 & 2 \\ -2 & 3 \end{pmatrix} = \begin{pmatrix} 1 & 1 \\ 7 & -4 \end{pmatrix}$

If the zero matrix $\begin{pmatrix} 0 & 0 \\ 0 & 0 \end{pmatrix}$ is added or subtracted no change is produced. This matrix is known as the additive identity.

Multiplication is more complicated; consecutive elements in the first row of the first matrix are multiplied with corresponding elements in the first column of the second matrix and the products of each pair of elements are added together to produce the element which is placed in the first row and the first column of the product matrix. This process is repeated with the second row and second column, and so on until the product matrix is calculated.

$$\begin{pmatrix} 1 & 2 \\ -2 & 3 \end{pmatrix}\begin{pmatrix} 2 & 3 \\ 5 & -1 \end{pmatrix} = \begin{pmatrix} 1 \times 2 + 2 \times 5 & 1 \times 3 + 2, \times -1 \\ -2 \times 2 + 3 \times 5 & -2 \times 3 + 3, \times -1 \end{pmatrix} = \begin{pmatrix} 12 & 1 \\ 11 & -9 \end{pmatrix}$$

If we reverse the order of multiplication

$$\begin{pmatrix} 2 & 3 \\ 5 & -1 \end{pmatrix}\begin{pmatrix} 1 & 2 \\ -2 & 3 \end{pmatrix} = \begin{pmatrix} 2 \times 1 + & 3 \times -2 & 2 \times 2 + & 3 \times 3 \\ 5 \times 1 + & -1 \times -2 & 5 \times 2 + & -1 \times 3 \end{pmatrix} = \begin{pmatrix} -4 & 13 \\ 7 & 7 \end{pmatrix}$$

Now $\begin{pmatrix} 1 & 2 \\ -2 & 3 \end{pmatrix}\begin{pmatrix} 2 & 3 \\ 5 & -1 \end{pmatrix} \neq \begin{pmatrix} 2 & 3 \\ 5 & -1 \end{pmatrix}\begin{pmatrix} 1 & 2 \\ -2 & 3 \end{pmatrix}$

So matrices, like Hamilton's quaternions, are not in general commutative; and there are two kinds of multiplication of one matrix by another, pre-multiplication and post-multiplication:

$\begin{pmatrix} 1 & 0 \\ 0 & 1 \end{pmatrix}$ is known as the unit matrix or the multiplicative identity. It produces the same result whether it pre or post multiplies another matrix.

$$\begin{pmatrix} 1 & 0 \\ 0 & 1 \end{pmatrix} \cdot \begin{pmatrix} 2 & 3 \\ 5 & -1 \end{pmatrix} = \begin{pmatrix} 2 & 3 \\ 5 & -1 \end{pmatrix}\begin{pmatrix} 1 & 0 \\ 0 & 1 \end{pmatrix} = \begin{pmatrix} 2 & 3 \\ 5 & -1 \end{pmatrix}$$

Inverses can also be calculated and, among other things, they can be used to solve systems of linear equations.

Matrices have many applications in geometry as well as in algebra.

The idea of a group, which was suggested by Galois, was extended by Cayley. He described the notion of an abstract group with a general structure which could be applied to numerous particular cases such as permutations, quaternions and matrices. Cayley showed that any particular group has the same structure as a subgroup of the symmetric group, the group of all permutations of a given order. The now familiar form of the group multiplication table was a device used by him, and is now sometimes called a Cayley table. A diagrammatic form of the structure of a group using connected vertices, with the vertices as the elements of the group, is also used extensively and is now called a Cayley diagram.

In the only group of order 3, the cyclic group of that order, with operation · and elements e, a, b, where e is the identity elements, the Cayley table and the Cayley diagram are:

·	e	a	b
e	e	a	b
a	a	b	e
b	b	e	a

Cayley table

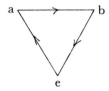

Cayley diagram

CHAPTER FIVE

Richard Dedekind

THE INTEGERS, the natural numbers taken together with a + or — sign and zero, can be represented as points on a line extending without limit in either direction.

$$-5\ -4\ -3\ -2\ -1\ \ 0\ +1\ +2\ +3\ +4\ +5\ +6$$

The integers can also be used to make up a set of numbers with integral numerators and denominators which includes the integers e.g. $\frac{+1}{1}$ or $\frac{-2}{1}$ and the common fractions e.g. $\frac{1}{2}$. All these can be shown on a line. For simplicity we will consider only the positive side of the line and omit the sign.

$$0\quad \frac{1}{8}\ \frac{1}{4}\ \frac{3}{8}\ \frac{1}{2}\quad \frac{3}{4}\quad \frac{1}{1}\quad \frac{3}{2}\quad \frac{2}{1}\quad \frac{5}{2}\quad \frac{3}{1}$$

With a little imagination, it is not difficult to see that we can show any number which can be expressed as a ratio of two integers on this line. Even exotic numbers like $\frac{1597}{1753}$ can be presented, providing we draw a diagram sufficiently large and take care in

Richard Dedekind (1831-1916) *by courtesy of the Library of the Technical University of Brunswick*

our measuring. All numbers of this type whether positive or negative, are known as rational numbers.

Two related questions follow. Are there irrational numbers, which cannot be expressed as the ratio of two integers? and is the line filled by these rational numbers? The answers are related but lead us in slightly different directions. The ancient Greeks answered the first question and, less directly, the second also with the famous theorem of Pythagoras.

The rule that the sum of the squares on two sides of a right angled triangle is equal to the square on the hypotenuse was not new at the time of Pythagoras. The Babylonians had used special cases of this much earlier and the Egyptians and the Indians had explored the so called Pythagorean triples, triples of numbers like (3,4,5) and (5,12,13) which numerically satisfy the relationship $a^2 + b^2 = c^2$, e.g. $3^2 + 4^2 = 5^2$ and $5^2 + 12^2 = 13^2$. Pythagoras spent much of his life travelling around the Middle East. It is known that he spent many years visiting the temples of Egypt; and he almost certainly collected the wisdom and 'magic' of the peoples he visited. Even though he did not discover the rule, it is still believed that he proved it as the theorem bears his name and that he sacrificed an ox in honour of his discovery. This proof is important for the development of numbers because while only the integers were used as in $3^2 + 4^2 = 5^2$ no difficulties arose, but when the theorem could be applied to any triangle results like $1^2 + 1^2 = 2$ were obtained. Here 2 is the length of the hypotenuse squared, so the length of the hypotenuse must be $\sqrt{2}$.

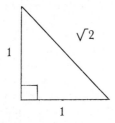

Clearly $\sqrt{2}$ is the length of a line and so can be represented on our number line, by construction with compasses at least.

(We will now omit the denominators for rational numbers equivalent to the integers.)

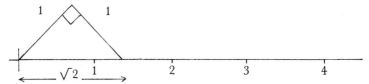

Now, is $\sqrt{2}$ a rational number? The only way to solve this problem is to find a way of expressing $\sqrt{2}$ with an integral numerator and denominator, or show that this is impossible.

The latter course was taken by Pythagoras, or his followers. Before we consider the proof, a mention can be made of the great man. Pythagoras was considered a rather magical character in his own time. He was thought to be a little above the mortals, but not quite a god. Many people were attracted to his society, which was organised rather like a commune. There were various levels of membership, some members lived an ordinary family life but acknowledged the society. However, the highest level of membership was very different from this. To belong to it a prospective member had to spend two years in silence somewhere near the master's feet, usually outside his tent. Later he was told the secrets of the society, but he was forbidden to tell the uninitiated. Those few who did commit this offence suffered various fates; the first perished by shipwreck but later banishment was used in which a grave would be made for the informant, as a mark of his death as far as the society was concerned, and he would neither be welcome nor acknowledged again. There was very much more to the Pythagoreans than the famous theorem, or even geometry; among many other things, they attributed magical properties to numbers. They lasted for several hundred years after the death of their master, who was murdered around 497 BC after a public uprising.

Because of the secrecy the Pythagoreans adopted, it is difficult to know exactly how they worked on the problems of the irrational numbers. Aristotle, the philosopher, gave a proof which has been attributed to them that $\sqrt{2}$ cannot be expressed as a ratio of two whole numbers. The proof used the method known as *reductio ad absurdum* (reduction to the absurd) and begins by supposing that $\sqrt{2}$ is rational. If this is so then $\frac{a}{b} = \sqrt{2}$, where $\frac{a}{b}$ is a rational number in its lowest terms. It follows that

$a^2 = 2b^2$ and that a is even, since the square of an even number is even.

Now let a = 2c
$$4c^2 = 2b^2$$
$$b^2 = 2c^2$$
and b must be even.

Now if a and b are both even then $\frac{a}{b}$ cannot be in its lowest terms. This is a contradiction of the original supposition, and therefore $\sqrt{2}$ cannot be expressed as a rational number $\frac{a}{b}$ and hence is irrational.

The Greeks were involved with $\sqrt{2}$ as a factor in the length of a line, in particular it was known from the theorem of Pythagoras that a square of unit length would have a diagonal of $\sqrt{2}$. They could not, however, find a common measure for both sides and diagonal of the square. With rational expressions a suitable unit measures both the denominator and the numerator exactly. No such unit could be found for both $\sqrt{2}$ and any natural number. Other numbers of this type are $\sqrt{3}$ and $\sqrt{5}$. Such numbers were said to be incommensurable, or to have no common measure with the natural numbers. The difficulties which incommensurables posed, contributed considerably to the stalemate which eventually enveloped Greek mathematics. As mathematics developed in more modern times the problems of irrational numbers and incommensurable magnitudes were gently ignored.

The same problem was to arise again when Richard Dedekind prepared a calculus course for his students and discovered that this aspect of the foundation on which his subject was based was very insecure. He then began to explore the nature and meaning of irrational numbers.

Julius Wilhelm Richard Dedekind was born in Braunsweig, Germany on 6th October 1831. His father, Julius, was a lawyer and a professer of law at the Collegium Carolinum in Braunsweig. His maternal grandfather was also a professor at the Carolinum. Richard, this was the only name he used later, was the youngest of four children. After some elementary schooling, he attended from the age of seven the Gymnasium Martino-Catherineum in Braunsweig. During the early part of his time at the Gymnasium his interests were centred around physics and chemistry. He saw mathematics as only an important tool in the sciences. However, as he went deeper and deeper into the study of physics he felt that the underlying principles and key experiments were not well founded. He also felt that physics lacked order. So if order was missing in the sciences, it was natural for him to turn to mathematics, and he had this in his mind when he left the Gymnasium in 1848 at the age of sixteen, and entered the Collegium Carolinum.

The Collegium, where his father and grandfather were professors, was academically on a higher level than the Gymnasium but not quite so high as a university. One of the greatest of all German mathematicians, Carl Friedrich Gauss, famous for his discoveries in many branches of mathematics and physics, especially the theory of numbers and electricity, had attended the same institute a number of years earlier. Richard was later to meet him in person. Dedekind's desire for mathematical knowledge was abundantly satisfied at the Collegium. He learned analytic geometry, algebra, calculus and mechanics, while continuing his studies in the sciences. Dedekind was well ahead of his classmates; and it is known that he gave private lessons to weaker or younger students during the second year of his studies.

After Easter in 1850, he entered the University of Göttingen. He had greatly benefited from his two years spent at the Collegium Carolinum; a course not normally taken by prospective university students who usually found the Gymnasium courses sufficiently demanding. As well as his more formal studies, he attended a seminar for future mathematics and physics teachers in Gymnasia. It was here that he met Bernhard Riemann, who was later to make revolutionary contributions to our knowledge of geometry. During his first semester, about half a year, he attended lectures on calculus (which just duplicated what he had already learned), hydraulics and experimental physics. He rarely attended laboratory sessions in physics but he was inspired by the lectures of Wilhelm Weber. In the rest of his studies he spent time on astronomy, and in later semesters he attended the lectures of Gauss on least squares and geodesy. These lectures were clearly remembered and admired by Dedekind, even in his later life.

After four semesters, Dedekind prepared for his doctorate, his supervisor was Gauss. Dedekind was, in fact, one of the last research students of the great man, who was by this time 75 years old. His thesis was on Eulerian integrals, a relatively remote branch of advanced calculus or analysis. Gauss' comments on his dissertation were that his work was independent and showed considerable knowledge, and that he had "favourable expectations of his future performance".

On graduating, Dedekind felt that his knowledge was sufficient for teaching in a Gymnasium, a type of employment frequently taken up by German mathematics graduates at this time, but was really insufficient in breadth for advanced studies and research. So he spent the next two years learning the new ideas of Steiner, Jacobi (remember Galois) and Dirichlet on the theory of numbers, advanced algebra, new developments in geometry and elliptic functions. In the summer of 1854 he was successful at Habilitation, which consisted of a thesis and an oral examination and qualified him to become a licensed, but unpaid, university lecturer or Privatdozent. To make a living in this position it was necessary to attract students to the courses offered, since the students paid the Privatdozent. Dedekind's lecture courses were on probability and geometry. In the latter he attempted to give good value for money by giving parallel treatments of the subject by both analytic and projective methods.

After the funeral of Gauss in 1855, where Dedekind was a pall bearer, Dirichlet came to Göttingen to fill his position. Dirichlet and Dedekind became good friends. Dirichlet's wife was the sister of Felix Mendelssohn the composer. At this time Dedekind was giving lectures and attending those of Bernhard Riemann. Riemann had joined the seminar for teachers years earlier with Dedekind and had Habilitated a few weeks before him. Of particular importance, is that during the winter semester of 1855-56 Dedekind gave what is probably the first ever university course on Galois theory. It is recorded that the lectures attracted few students, and only two were present when he introduced his idea of the abstract group, in place of Galois' notion of a group of permutations.

In 1858 he was called to become Extraordinarius or assistant professor at the Polytechnikum in Zurich (it is now the Eidenossische Technische Hochschule), largely on the advice of Karl Weirstrass. We shall return to this period, since it is important to the development of the theory of the irrationals, but first some more biographical information about Dedekind. Four years after he left Göttingen and twelve years after he left the Collegium Carolinum in Braunsweig he came full circle and was appointed Ordinariat or full professor at the Polytechnum in Braunsweig, which was the Carolinum under its new name. He had followed in his father's and grandfather's footsteps. He remained in this post until he retired and in his birthplace until his death. At the Polytechnum he felt it was important to undertake administrative responsibility, as his father had done. During the College's transformation to a Technical university, he was chairman of the building committee.

Richard Dedekind never married, and lived with his sister Julie who was a novelist of repute, for the rest of his life. He had many friends and relations in Braunsweig and felt happy and comfortable there. For relaxation he read novels, such as those of Sir Walter Scott, and played the piano and cello with considerable skill and feeling. Also he was talented as a composer; his magnum opus was a chamber opera written to the libretto of his brother Adolph.

When Dedekind began his course on differential calculus in Zurich, it was the first time he had looked at the subject as a teacher. To show how some of the underlying theorems were proved he used graphical or geometrical methods. Such methods imply that all the numbers used in the calculus can be identified with the points of the line, this is known as the Cantor-Dedekind axiom. This is really what we have used earlier and it brings us back to our earlier questions: is the line filled with numbers (or are there gaps), and can we represent every number by a point on the line? The fact that these questions could be raised was sufficient to cast doubt on the methods he was using to justify the calculus. If the answers to both questions were affirmative, then Dedekind's doubts could be vindicated. It is also important that the difficulties should be resolved by arithmetic, rather than by the traditional recourse to the apparent smoothness and continuity of the line representing a segment of the set of real numbers. It is true that numbers like $\sqrt{2}$, $\sqrt{3}$ and $\sqrt{5}$ could be shown to be points on a line by constructing their lengths with compasses and straight edge as allowed by the Greeks. However, this was not very satisfactory for arithmetic, as it was known that there existed other numbers, which defied such constructions i.e. transcendental numbers. The existence of these numbers was proved in 1844, while Dedekind was still at school, by Liuville. π, the ratio of the circumference of a circle to its diameter seemed likely to be transcendental since no-one had been able to 'square the circle', or construct with compasses and straight edge a square of the same area as a given circle. The transcendence of π was established by Lindemann in 1882. Another number, e, the base of natural logarithms, was shown to be transcendental by Hermite in 1873.

To attempt to solve his problems, Dedekind examined certain properties of numbers, in particular, the natural numbers and the integers form rather sparse sets, where it is not possible to insert another integer or natural number between any two others, except under special circumstances. For instance, we cannot insert another natural number between 1 and 2, but we can between 2 and 4 provided it is the mid-point as defined by the number line, in which case it is 3, but there is no other possibility. The rational numbers however, do not behave like this. Anywhere between any two natural numbers we can find another natural number. Consider that natural numbers $\frac{1}{1}$ and $\frac{2}{1}$ as represented on a line; by continually halving the distance and taking the lower number represented we get: $1, \frac{3}{2}, \frac{5}{4}, \frac{9}{8}, \frac{17}{16}, \frac{33}{32} \ldots$
This process can be repeated indefinitely because of the density of the rational numbers, but nowhere is there any possibility of an irrational number entering into the sequence. So the fact that the number line is very dense in many places does not get rid of the gaps. Very different ideas were needed to produce irrational numbers.

The idea that occurred to Dedekind was the very opposite of the notion of density. It was that, in his own words, "If all points of the straight line fall into two [sets] such that every point of the first set lies to the left of every point of the second set then there exists one and only one point which produces this division of all points into two sets, thus severing of the straight line into two portions". [Dedekind uses 'class' instead of 'set']. In other words, for any number, a, on the number line there is a set of numbers A_1 to the left of it such that every number it contains is less than a and a set of numbers A_2 to the right of it such that it contains every number greater than a.

$$A_1(<a)\ a\quad A_2(>a)$$

This separation of the number line Dedekind called a cut (Schnitt in German) and showed it as (A_1, A_2). This cut can be produced by a rational number or by an irrational number. For example, $\sqrt{2}$ is defined by the Dedekind cut where A_1 is the set of all negative rational numbers and all positive rational numbers with squares less than 2 and A_2 is the set of all positive rational numbers with squares greater than 2. So purely by an arithmetical process it is possible to get an irrational number from the set of rational numbers.

This discovery was recorded in one of Dedekind's notebooks, and is dated 24th November, 1858. He did not have a great deal of confidence in his idea at first, but the subsequent development of mathematics in this area suggested its worth and he finally published it in 1872 in *Continuity and Irrational Numbers*. In this work he mentioned one of the great difficulties of irrational numbers, which is important at school. Using his own example, $6 = 2 \times 3$ and so $\sqrt{6} = \sqrt{2 \times 3}$ but he knew of no way to prove the next step that $\sqrt{6} = \sqrt{2} \times \sqrt{3}$. This is something which was taken for granted, as any arithmetical calculation would point towards its truth, but only approximately. He believed that the idea of the 'Dedekind cut' would give a final proof.

His early lack of confidence in his construction of irrational numbers was not unfounded as it came in for considerable criticism. The English mathematician, logician and philosopher, Bertrand Russell suggested that he had assumed the existence of such numbers in order to construct them. Russell also said that only one set either A_1 or A_2 was necessary, since the existence of A_1 implied the existence of A_2 and vice versa. He goes on to say in his *Introduction to Mathematical Philosophy* that either one of

Dedekind's sets is sufficient for the definition of an irrational number, rather than the cut itself, and that such a number is defined, for example, by the ordered set which is less than $\sqrt{2}$. An irrational number was thus an order property of the set of rational numbers.

Other definitions of irrational numbers were given by Meray, Cantor, Weierstrass and Heine. Indeed, it was the work of Heine which prompted Dedekind to publish his work. Kronecker completely denied the existence of irrational numbers and believed that arithmetic should be constructed without them. Also, whether or not Russell's definition is superior to that of Dedekind, it is that of the latter which is used now.

The conquest of the irrationals was not Dedekind's only research with numbers, his greatest work was yet to come. This was concerned with the theory of numbers and his results were published in 1871 as a supplement to Dirichlet's book on number theory *Vorlesurgen über Zahlentheorie*, which Dedekind edited. Subsequent developments in his work were published as supplements to later editions of the same book. Although in general this part of Dedekind's work is too advanced to be discussed here, there are some points which are both interesting and sufficiently elementary to be described.

Again the story begins with the ancient Greeks and in particular with Euclid. One of the most beautiful results in the whole of mathematics is that any integer, except zero, can be represented uniquely as (± 1) times a product of prime numbers. For example, $9590 = 2 \times 5 \times 7 \times 137$. The order of the prime numbers in the representation is unimportant. This is Proposition 14 of Book VII of *Euclid's Elements*.

C. F. Gauss, who had been Dedekind's teacher, was responsible for showing that there was a similar theorem for complex integers; that complex numbers could be uniquely factorised as a product of complex prime numbers.

Other mathematicians explained the possibility of unique factorisation in other sets of numbers and gave information about the conditions under which this factorisation could occur.

If we take a familiar starting point, $\sqrt{2}$ and $-\sqrt{2}$ are the roots, or solutions for x, of the quadratic equation $x^2 - 2 = 0$. Now consider the whole class of equations of which this quadratic equation is a simple example, that is equations of the form

$a_1x^n + a_2x^{n-1} + a_2x^{n-2} + \ldots + a_n = 0$. When n = 1 we obtain simple examples such as x − 2 = 0, when n = 2, we obtain quadratic equations, when n = 3 we obtain cubic equations and so on. Provided that the coefficients $a_1, a_2,\ldots$ are rational numbers, the roots of such equations are known as algebraic numbers. $\sqrt{2}$ is an algebraic number of degree 2 (since n = 2 in $x^2 - 2 = 0$), the rational numbers themselves are algebraic numbers of degree 1 (since n = 1), $\sqrt[3]{2}$ is an algebraic number of degree 3 and so on. When the coefficient of the highest power of x in the equation is 1, $(a_n = 1)$ the roots are known as algebraic integers. Certain arithmetics based on particular subsets (fields) of the algebraic numbers have very strange properties when we examine them for unique factorisation.

Consider the equation $x^2 - 4x + 9 = 0$. By using the familiar formula $x = \dfrac{-b \pm \sqrt{b^2 - 4ac}}{2a}$ for the equation $ax^2 + bx + c = 0$ we find that $x = \dfrac{4 \pm \sqrt{16 - 36}}{2}$ or $x = 2 + \sqrt{-5}, 2 - \sqrt{-5}$. Thus $2 + \sqrt{-5}$ and $2 - \sqrt{-5}$ are algebraic numbers of degree 2. Now $(2 + \sqrt{-5})(2 - \sqrt{-5}) = 9$, but also $3 \times 3 = 9$. So unique factorisation fails in an arithmetic which includes the algebraic numbers 3, $2 + \sqrt{-5}$ and $2 - \sqrt{-5}$, and multiplication. The mathematician Kummer, thought he had a way of both resolving the problem and preserving unique factorisation. He devised 'ideal numbers' p and q such that:

$$9 = (pq)^2, 3 = pq, 2 + \sqrt{-5} = p^2, 2 - \sqrt{-5} = q^2.$$

So that $9 = (pq)^2 = 3^2 = p^2q^2 = (2 + \sqrt{-5})(2 - \sqrt{-5})$

Unfortunately Kummer's 'ideal numbers' were neither algebraic numbers, nor were they defined in a way suitable for the construction of the whole set.

Dedekind explored this problem and developed what he called 'ideals', the name 'ideal' was used in honour of Kummer. To show how he tackled the problem we can look at another example, this time not involving algebraic numbers but using a set which possesses similar difficulties with regard to unique factorisation. This example was constructed by David Hilbert, whose life and work we shall examine in detail later. The set is that of the numbers which leave a remainder of 1 when divided by 4. Technically this is known as the set of positive integers congruent to 1 modulo 4 and is:

$$\{1, 5, 9, 13, 17, 21, 25, 29, 33, 37, 41, 45, 49,..., 441,...\}.$$

For these numbers our definition of a prime number, is a number which cannot be decomposed into factors which are also elements of the set. So 5, 9, 13, 17, 21, 29, 33, 37, 41, 49... are prime numbers and $25 = 5 \times 5$ and $45 = 9 \times 5$ are not. Now $441 = 21 \times 21$ and $441 = 9 \times 49$ and hence can be expressed in primes in two ways. A little thought will show that if 3 and 7 had belonged to our set then unique factorisation would have been possible. It is also clear that 21 and 9 are related, because their highest common factor (H.C.F.) is 3, and similarly 21 and 49 are related with H.C.F. 7. If we could replace the offending prime numbers by suitable H.C.F.'s. then the problem would be solved. That is, if $441 = (21,9)^2 (21,49)^2$, where (21.9) represents the H.C.F. of 21 and 9 and (21.49) represents the H.C.F. of 21 and 49. Notice the numbers 3 and 7 do not themselves enter into the argument, so we are not justified in writing $441 = 3^2 \times 7^2$; it is the properties of the H.C.F.'s that are used, not their values. It is these properties which Dedekind examined, and from which he developed his concept of the 'ideal'. Ideals are, in essence, generalisations of integers, which allow the algebraic numbers the property of unique factorisation.

Dedekind introduced the terms 'field' and 'ring' into the vocabulary of mathematics. They were originally used by him in connection with his ideals, but now they are used in many more branches of mathematics. A 'field' is the set of all the numbers such that if a and b belong to the set then so does $a + b$, $a - b$, ab and, provided $b = 0$, $\frac{a}{b}$. A 'ring' is defined in the same way as a field, except that $\frac{a}{b}$ is not required to belong to it.

Dedekind was recognised as one of the greatest mathematicians of his time, yet he stayed in one of the less prestigious of the German universities for most of his life. Why he did not join the merry-go-round of professorial appointments is not known, but, perhaps that he was with his family and friends in his home town has something to do with the answer.

Among his friends was Georg Cantor, whom we will meet in the next chapter. They spent some holidays together at Interlaken and had various periods of correspondence to the mutual development of their research. When a mathematical publication reported Dedekind as having died on the 4th September 1899, he was greatly amused because he passed the day in perfect health, having had lunch and a vigorous discussion with Cantor.

In 1914 his sister Julie, with whom he had lived for more than fifty years, died. Two years later, in 1916, the still alert Dedekind, passed away at the age of 85.

CHAPTER SIX

Georg Cantor

THE NOTION OF A set is a common one. We speak of a "set of tools", a "train set", a "twin set" and there are many more examples of the word "set" in our everyday language. It means simply, a collection of objects which usually, but not always, have a common purpose or connection. The idea of a set in mathematics has been used implicitly since the time of the ancient Greeks, or even before. Now their use and the algebra associated with them is common in school mathematics. One person, Georg Cantor, is largely responsible for developing this concept from a mere descriptive notion into a powerful tool which is useful from the mathematics of the infant school, to that at the frontiers of research.

Georg Ferdinand Ludwig Philipp Cantor was born on the 3rd March, 1845, in St. Petersburg, where his father was a merchant. This town was also the birthplace of his mother, but his father was born in Copenhagen, having come to Russia to carry on his business. His father Georg Woldemar Cantor was of Jewish extraction, but had been brought up in a Lutheran mission and was a staunch adherent to the protestant faith. His mother Maria Anna Bolm came from a family which had shown prodigious musical ability and artistic talent. Georg was the first of their six children.

Georg was first taught reading and writing by his mother, and the Christian faith by his father. Later he went to an elementary school in St. Petersburg. In 1856 the family moved to Wiesbaden, and then settled in Frankfurt where Georg went to private schools. He then went on to the Wiesbaden Gymnasium

Georg Cantor (1845-1918) *by courtesy of the Archives of the Martin Luther University, Halle-Wittenberg, German Democratic Republic*

and in 1859 to the Grand-Ducal Realschule in Darmstadt, where he was a boarder. The following year he attended the Hoheren Gewerbschule (trade school), where he stayed until 1862.

From the beginning of his schooling Georg had shown himself to be a person of many talents and high ability. He had excelled in mathematics, and from being quite young he had felt himself drawn towards the study of this subject.

Throughout his childhood he was encouraged in his studies and in many other ways by his parents. The discipline of the household was fairly strict. There was always a striving for success and accomplishment. The Christian faith, in the arduous Lutheran form, pervaded the whole family. Cantor survived and thrived in this environment. He developed an interest in theology and philosophy. Artistic encouragement from his mother gave him a deep and scholarly love of literature. The hours of musical practice made it possible for him to play in a string quartet when his parents put up other musicians for a Christmas holiday. He also showed a flair for drawing and painting. At school, as a boarder, there was less sympathy and empathy, but the rigid discipline of German education served only to develop his gifts further. The only subjects in which he did not consistently excel were geography and history.

All was not completely well. There was a driving force from the family to succeed, and succeed he did, but it is believed that he suffered even then from the fits of depression which were to play such a great part in his later life. Letters from his father say that he would soon cheer him up and make him snap out of the doldrums into which he had sunk when he came home at the end of term.

With the best of motives, his father wished him to use his mathematical and scientific talents to advantage by training as an engineer. From the elder Cantor's point of view, this was the obvious career. He had succeeded in business (it is believed that on his death he left over half a million marks), and engineering was important to trade. Engineers were needed and, because they were scarce, highly paid. Georg did not want this, and eventually his father showed his sensitivity and wisdom by allowing him to proceed to university to study mathematics as his principal subject. However, he advised him to keep his studies as general as possible. He need not have worried, Georg became an expert in the fields of theology, philosophy and literature as well as mathematics.

In 1862 Cantor began his higher education at the Polytechnikum in Zurich. He again experienced periods of depression. The following year his father died from tuberculosis and, after the family's affairs were put in order, Cantor went to Berlin. Freed now from his father's promptings, he probably chose the university at the capital because of its fine mathematics professors. The prestige of Berlin was tremendous since it was the pinnacle of achievement for university professors, who tended to move from the smaller universities to the larger ones, at the invitation of the German Ministry of Culture.

The senior members of the mathematics department at Berlin were Weierstrass, Kummer and Kronecker. There were relatively few students of mathematics so they enjoyed a close relationship with their professors.

Although it was common for German students to travel from one university to another sampling the academic wares of the various universities, Cantor spent only one period away from Berlin, the summer of 1866 in Göttingen.

In 1867 he was granted his doctorate from a thesis on the theory of numbers and the theory of indeterminate equations applied to a point bypassed by C. F. Gauss in his famous work on the theory of numbers. Although the twenty-six pages of the thesis were sufficient to gain him the distinction of Magna Cum Laude (high honours) they showed little of the revolutionary inspiration that was to be the hall mark of his later work. The following year he passed the Staatsprüfung, the examination which intending teachers in state secondary schools were required to pass. He then spent a short time as a teacher in a girl's school, but carried on his research into the theory of numbers at the same time.

Early in 1869 he went to the university in Halle as a Privatdozent. The meagre income Cantor earned came from his students, but he was generally independent because of his father's business fortune. Halle was a distinctly lower grade university. Nevertheless, it was a suitable place for an aspiring young academic to begin. During these early years as a university teacher he turned away from his work on numbers and carried out research on analysis, a very advanced version of calculus. In particular, he investigated series made up from the sines and cosines which we encounter in trigonometry. Various forms of these are mathematically interesting and have important applications in physics and engineering. The pioneer of these was Joseph Fourier, after whom certain kinds of trigonometric series are named. As Cantor's success with series and as a university lecturer grew, he was promoted to Extraordinarius in 1872 and to Ordinariat in 1879. He was, however, very conscious that Halle was one of the lesser German universities. There were very few students pursuing research and no able and aspiring young mathematicians wishing to join the mathematics department as staff members. Cantor naturally hoped that one day a call would come to take up a senior position with one of the more prestigious universities, Berlin being preferred.

The discoveries which should have ensured his academic success arose fairly naturally from his work on trigonometric series. Collections of infinite numbers of points began to appear as an important aspect of the series he was studying. The idea of infinite numbers, the nature of infinity and the collections which we now call sets, became the focus of his work. A new theory of the infinite was needed; and this as both a mathematical and theological idea intrigued Cantor. His development of the theory of sets is now the most commonplace part of his work for many of us, but this was initially an idea he developed to handle his idea of infinite numbers.

Before we examine Cantor's ideas we need to explore the history of the notion of infinity and some of the strange things connected with it.

In the *Physics* book by the ancient Greek philosopher Aristotle written three and a half centuries before the birth of Christ, the four paradoxes of Zeno the Eleatic, which deal with motion, are described. Zeno lived about a hundred years before Aristotle, and during his time the philosophers held two opposing views of space and time. Either that they were infinitely divisible, so that motion would be smooth and continuous, or that they were made up from tiny indivisible parts. The four paradoxes, taken together, showed that neither argument made complete sense.

The first paradox, known as the Dichotomy says that motion is impossible since something travelling must first reach the half way point (A_1), but before that it must have travelled half this distance (A_2), and before this it must have travelled half the previous distance (A_3) and so on, therefore the goal can never be reached.

Start ├─┼─┼─┼────────┼─────────┼───────────────────┤ Goal
       $A_5 A_4\ A_3$ A_2 A_1

The second one is the Tortoise and Achilles, where the slowest moving object (the Tortoise) given a head start, cannot be passed by the quickest (Achilles), because as Achilles reaches the point from which the Tortoise started, it has moved a little further and by the time Achilles has reached the new position the Tortoise has again advanced, and so on.

These two paradoxes say that motion and relative motion are impossible when space and time are infinitely divisible. The next two, the Arrow and the Stadium suggest that motion is impossible if space and time are made up from infinitely small indivisible elements. The Stadium deals with relative motion and is difficult to visualise, but the Arrow explains that if time is made up from instants, then at each instant the arrow occupies a definite position element, and since space is not in motion the Arrow is at rest.

There is obviously something wrong with the arguments of Zeno because they contradict our everyday experience. They use two different kinds of infinity. One is an infinite process of dividing, the other is an infinite number of parts. Aristotle describes both of these and uses them to explain Zeno's paradoxes. Because there is something very strange about these ideas of infinity and because the works of Aristotle was one of the main sources of accepted knowledge, the explanation of 'infinity' was left alone for two thousand years.

The next really important person to become involved was Galileo. In his *The Two New Sciences*, published in 1638, he explains how the points on two lines can be made to correspond exactly.

In the diagram AB and CD are the two unequal lines, suppose that AB is 7cm and CD is 4cm. Wherever we draw a line from point O to a point P on AB the line passes through a point P_1 on CD. So there are as many points on the line CD (4cm) as there are on the line AB (7cm).

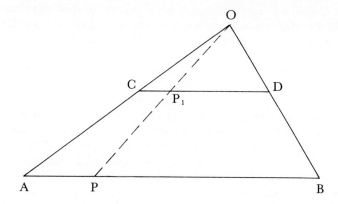

Galileo also noticed that we can make the natural numbers correspond with their own cubes.

$$
\begin{array}{ccccccc}
1 & 2 & 3 & 4 & 5 & 6 & 7 & \ldots \\
\downarrow & \downarrow & \downarrow & \downarrow & \downarrow & \downarrow & \downarrow \\
1^3 & 2^3 & 3^3 & 4^3 & 5^3 & 6^3 & 7^3 & \ldots
\end{array}
$$

This suggests that there are the same number of natural numbers as there are cubes of the natural numbers. Is this a contradiction? Galileo thought this was the case and rejected the idea of infinities of numbers.

As differential calculus developed it became necessary to deal with infinite processes in which limits were involved, but the great mathematicians, in particular Gauss and Cauchy, emphasised that these limits were not numbers. They were not really anything which could be properly described, nor were they complete in themselves, they were only a means of explaining the process of increasing without restraint.

When Georg Cantor began to explore the nature of infinity, to come up with any conclusion other than that it was a vague mathematical monster that should be avoided whenever possible was a quick way to mathematical suicide. This was very nearly what happened. Like many revolutionary ideas, Cantor's notion of infinity was treated with incredulity, then with a near conspiracy to stop it spreading further and lastly with an attempt by the mathematical community to close ranks and ignore it.

Cantor's first revolutionary discovery was to show that there are as many points in the whole of a plane as there are in a line. This seems to extend Galileo's discoveries that the number of points in a line does not depend on its length. In the chapter on Peano we shall meet the consequence of this fact, and a further confirmation of its truth. This discovery was looked upon with either extreme caution or plain disbelief by many of the established mathematicians.

Leopold Kronecker, one of Cantor's teachers in Berlin and an independently wealthy man as well as an academic, was the most incredulous of them all. A well known quotation of his was that "God created the natural numbers, man created the rest". Kronecker was working on a programme of research designed to prove this statement by showing that all the types of numbers which we know can be obtained mathematically from His divinely created natural numbers. He had a vested interest against the discoveries of Cantor since these, if correct, would demolish many of the foundations upon which he based his research. Kronecker reacted strongly and seriously criticised Cantor's results and methods, and generally let it be known that he felt Cantor was at best an inferior mathematician and at worst not a mathematician at all.

The results of Cantor's examinations of infinite sets of numbers were published from about 1874 onwards with the main statements being made in 1878, 1883 and particularly in 1895 and 1897. It will help us to consider his ideas if we use the formulations that appeared in his later papers.

The first sentence of the 1895 work contains the well known definition of a set: 'By a "set" we mean any collection M into a whole of definite, separate objects m of our intuition or our thought. These objects are called the "elements" of M'. He first called the number of elements in a set its power, but later changed this to its cardinal number. Sets with the same number of elements have the same cardinal number.

The set of numbers which is basic to all arithmetic is the set of natural numbers.

$$\{1, 2, 3, 4, 5,...\}$$

This is the infinite set which puzzled Galileo; but Cantor was more bold, he assigned a definite cardinal number to it. The number was $\aleph_0$ aleph null, aleph being the first letter of the Hebrew alphabet. The choice of this symbol for the first infinite or transfinite cardinal number was not idle. For his new concept, Cantor wanted a symbol which had not taken any other mathematical role. It is also used as 'one' in Hebrew. Cantor believed that there was a whole series of transfinite cardinals, $\aleph_0, \aleph_1, \aleph_2 \ldots$

Any set where elements could be paired off, or put into one-to-one correspondence, with the elements of the set of natural numbers was said by Cantor to be denumerable. Galileo had shown that the set of cubes of the natural numbers is denumerable. With a little imagination, or a pencil and paper, it will not be difficult to see that the sets of even numbers, odd numbers and squares of the natural numbers are denumerable. With a little trickery it is possible to show that the set of integers $\{\ldots -2, -1, 0, 1, 2, \ldots\}$ is also denumerable. In fact, one criterion for a denumerable set is that a proper subset of it can be put into one-to-one correspondence with the set itself.

The set of points in a line, or the set of real numbers which are considered to be the points in a line, did not seem to be the same as the set of natural numbers. But to prove this was a different thing. Cantor achieved this in two ways, but his later version is the simplest.

If we consider a real number as being made up from an unlimited number of decimal places, and if we take as many real numbers as we like, as close together as we like, we can always construct another one. For instance, from the numbers

$$0. a_1 a_2 a_3 a_4 \ldots$$
$$0. b_1 b_2 b_3 b_4 \ldots$$
$$0. c_1 c_2 c_3 c_4 \ldots$$
$$0. z_1 z_2 z_3 z_4 \ldots$$

he produced a new one which was not a_1 in the first decimal place, not b_2 in the second place and so on until it differed from each of the other numbers in at least one decimal place.

New number 0. $a'_1 \, 'b'_2 \, c'_3 \ldots z'_i$ (where $'$ means 'not').

By means of this diagonal proof he was able to generate a new number from any given subset of the real numbers. This was impossible with the natural numbers. To the set of real numbers he assigned the cardinal number c.

Cantor explored other sets of numbers to discover whether they were denumerable, with cardinal number $\aleph_0$. One discovery, which lowered him further in the eyes of Kronecker was that the set of algebraic numbers (the numbers which can occur as the solutions of polynomial equations) is denumerable. He found that the infinite sets of numbers which seemed to arise normally in mathematics had either $\aleph_0$, or c as their cardinal numbers.

The set of natural numbers, or any denumerable set, can be ordered. It can have first, second, third and so on elements. Cantor went on to devise transfinite ordinal numbers. He suggested that the set of natural numbers had elements $1, 2, 3,...\omega$ where ω was the first transfinite ordinal. This was the previously forbidden actual, or completed, infinity. ω, the last letter of the Greek alphabet was purposely used to indicate an element at the end of the finite natural numbers. There were, however, other transfinite ordinal numbers, $\omega + 1, \omega + 2,..., 2\omega ... \omega^2 ...$; this set has cardinal number $\aleph_1$. Had he investigated further, he might well have discovered transfinite prime numbers!

From his examination of the properties of sets he was able to show, with a very simple proof, that the set of fractions or rational numbers was denumerable. Its elements could be ordered by adding their numerators and denominators together and keeping all those with the same sum together and in order by their numerators:

$\underline{1}, \underline{1}, \underline{2}, \underline{1}, \underline{2}, \underline{3}, \underline{1}, \underline{2}, \underline{3}, \underline{4}, \ \underline{1}, \ \underline{2}, \ \underline{3}, \ \underline{4}, \ \underline{5}, ...$
$1 \ 2 \ 1 \ 3 \ 2 \ 1 \ 4 \ 3 \ 2 \ 1 \ \ 5 \ \ 4 \ \ 3 \ \ 2 \ \ 1$
↕ ↕ ↕ ↕ ↕ ↕ ↕ ↕ ↕ ↕ ↕ ↕ ↕ ↕ ↕
$1, 2, 3, 4, 5, 6, 7, 8, 9, 10, 11, 12, 13, 14, 15 ...$

The problem of the relationship between the two alephs $\aleph_0$ and $\aleph_1$ was a cause of concern for Cantor right up to the end of his days. From a set $\{a, b\}$ we can get the set of its subsets, including the set itself and the empty set (the set with no elements), $\{ \{ \}, \{a,\}, \{b,\}, \{a, b,\} \}$ known as the power set. The cardinal number of this power set is 4 or 2^2. The cardinal number of the power set with three elements is 8 or 2^3, that of the set with four elements is 16 or 2^4 and, in general, the cardinal number of the power set with n elements is 2^n. Is the set of real numbers the set of all subsets which can be made from the natural numbers? If this is the case it would have $2^{\aleph_0}$ elements and c would equal $2^{\aleph_0}$. Cantor believed that $c = \aleph_1$, his second transfinite cardinal, and since the real number line, the continuum, has the same number of points as the set of real numbers, this was known as his continuum hypothesis. Many times he believed he had proved this, only to find a flaw in the proof. He never solved the problem, and as we shall see in a later chapter it was eventually considered to be of paramount importance by David Hilbert.

Although Cantor mainly examined infinite sets, he devised most of the ideas which we use in the finite sets common in school mathematics. From his explorations of transfinite numbers we have gained a precise concept with phenomenal power and ubiquitous manifestations in finite, as well as transfinite mathematics.

Kronecker used his influence as one of the foremost German mathematicians to try and suppress Cantor's work by active means, as well as by trying to show that it was ridiculous and not worthy of the name mathematics. In 1877 Cantor attempted to publish some of his research in a journal which had Kronecker as one of its advisory board. The other members agreed to its publication, but after a long time the paper was still not being prepared for publication. Only after Cantor let it be known that he would withdraw it and publish it privately as a pamphlet, was it published.

Dedekind, who, as we have seen, was subject to second rate treatment himself by the German mathematical establishment was one of the few who befriended Cantor. After their wedding in 1874, Cantor and Valle Guttmann spent their honeymoon in the Hartz mountains and during some of these summer days Cantor discussed his ideas with Dedekind. There was a considerable correspondence between them over the irrational numbers as well as the transfinite numbers and the theory of sets.

Another friend of his had been Mittag-Lefftler, who as editor of the influential journal *Acta Mathematica* had considered his articles favourably and given him, by their publication, a means of presenting his discoveries to the mathematical community when other journals treated him unsympathetically. But in 1885, when Cantor tried to publish two short articles in *Acta Mathematica*, Mittag-Leffler wrote suggesting that his work was not yet fully developed for publication. Cantor then believed that even this friend had been influenced against him and was now only interested in protecting the reputation of his journal. After this incident he submitted his research to other journals, but apart from his two classic papers of 1895 and 1897, he published very little during the rest of his life.

Throughout his time as a mathematics professor, Cantor had hoped that he would be 'called' to one of the more important universities. At Halle, where very few students stayed to take their doctorates and able mathematics lecturers usually moved on with promotion, he had little opportunity for influencing either the senior mathematicians or the aspiring research students from other universities. Cantor would have loved to move to Berlin or Göttingen but no call came, not even one to a slightly less distinguished mathematics department. Indeed, when he knew that a position was due to become vacant in Berlin he wrote directly to the minister responsible for such appointments asking to be considered. However, as he suspected, he did not receive the desired invitation. He believed that Kronecker used his influence to thwart his aspirations.

In the early part of October 1884 Cantor spent an evening with Kronecker at his home in an attempt to reconcile their differences. The evening was not the disaster it could have been but nothing changed and Cantor continued to be made miserable by his one time mentor until the latter's death in 1891.

From about 1884 several circumstances conspired to disillusion Cantor. He had found no success with his continuum hypothesis, every promising approach was discovered to contain a flaw. He had been snubbed, or so he thought, by Mittag-Leffler. The recognition that was rightly his was not forthcoming from either his fellow mathematicians or the German education authority. He had also suffered the first of his serious mental breakdowns and had spent some time in hospital. At this point he turned away almost completely from mathematics. During his early life he had many interests, art, music, theology, philosophy and literature. There was, perhaps, some regret during this and later periods of his life that he had not chosen another field for his life's work. His interests now centred upon theology and, particularly, philosophy, since this was a subject in which he was well read and of a certain standing. Cantor began to teach philosophy, and to correspond with philosophers and theologians over the interpretation of his ideas of the infinite. During the early part of 1884 he had written to Mittag-Leffler saying that God was the real centre of his work; he was only the means of writing it down.

These were not the only areas in which he had interests. He developed interests in freemasonry, the secret religious society of the Rosicrucians and the expanding theosophy movement. There was also another area in which he was very knowledgeable and to which he devoted time in research and lecturing. It is sometimes believed that the works of William Shakespeare were not written by him at all, but by any number of his contemporary playwrights, one of these being Francis Bacon. There is a certain amount of evidence to support Bacon's claim to the authorship, and a controversy has existed over the years concerning the evidence for either contender, Bacon or Shakespeare. Cantor was very knowledgeable in this area. He read the history and documents concerning the Elizabethan playwrights with the hope of showing that Bacon was the rightful author. It was after returning home from delivering a lecture on this question in Leipzig that he learned of the death of his musically gifted son Rudolph.

During his later life hs suffered increasingly from mental depression. He had a number of breakdowns, increasing in length and severity, some of which needed treatment in hospital. During 1899, while he was in a poor mental state, he applied to the authorities for any position that would give him his present salary but would take him away from the university. His continuum hypothesis, to which he often returned, continued to elude all his intellectual efforts. His inability to solve this problem must have made at least a contribution to his mental distress. Between 1904 and 1905 he was in hospital for nearly a year, and spoke and wrote about hearing a divine voice which guided his study of theology.

Unfortunately just as his health was deteriorating, the mathematicians of the world were beginning to appreciate his achievements. Those who had criticised or ignored his work were being forced to take notice of it. He was awarded the honours he deserved, including medals and prizes from foreign academies.

Cantor never received the academic 'call' he so earnestly desired and spent most of the rest of his life associated with the university of Halle. As the years passed he suffered more and more from his mental illness and spent much of his time in hospital, where he died on 6th January 1918.

CHAPTER SEVEN

Felix Klein

GEOMETRY IS A FAR broader subject than that taught in most school courses. The geometry of Euclid, who was the collector and elaborator of most Greek geometry, is concerned with distances and angles, and many other ideas which can be derived from these. Co-ordinate geometry, at least in its more simple form, is the addition of algebraic methods to Euclidean geometry, together with an explicit concept of position. If only ideas more general than angle and distance (such as parallelism or the intersection of lines), are considered, then new geometries may result. In fact school geometry is a very restricted special case of the study of space.

Throughout the first half of the nineteenth century the mathematical exploration of the ideas of space proceeded at a rapid rate. Great strides were made by Gauss and Riemann, both of Göttingen. A whole new approach to the geometry of Euclid was made by Lobachevsky (1829), J. Bolyai (1832) and Bernard Riemann when they developed non-Euclidean geometry, which is similar to the Greek subject in all aspects but one. There was, in fact, a vast accumulation of deep and important geometrical knowledge without a systematic mathematical relationship connecting the various areas.

As will be remembered from the first chapter, at the beginning of the nineteenth century the group concept was being seen as an important idea in algebraic research by Evariste Galois. Although Galois' ideas concerning equations of the fifth degree took many years to be appreciated, the notion of the group, which was known to others besides him, was developed, extended

Felix Klein (1849-1925) *Bildarchiv Preussischer Kulturbesitz*

and refined and applied to various areas of mathematics. The bringing together of the various geometries, new and old, with the group idea was to result in a coherent structure of nearly all our knowledge of space. This synthesis of algebraic and geometrical structures was largely accomplished by Christian Felix Klein.

Felix was born in Düsseldorf, Germany on 25th April, 1849, just as the turbulence in the German States was beginning to settle. His father was a civil servant, a senior personal secretary. Felix attended schools in Düsseldorf, finally graduating from the Gymnasium in the summer of 1865. He entered the university of Bonn in the winter semester of the same year and specialised in mathematics and physics.

At first his love had been physics, and while still a student he had assisted his professor, J. Plucker, with his physics lectures. Plucker was then pursuing geometrical research, rather than physical, and as his assistant, Klein became deeply involved with this also. In 1868, the same year as the nineteen year old Klein gained his doctorate, Plucker died and Klein was asked to complete the second part of his mentor's book on line geometry.

The geometry we normally think of uses points as its basic element. However, a geometry can be developed which uses lines for the representation of space. There are similarities between the two kinds of geometry, for example, two lines define a point and two points define a line. These two definitions play equivalent roles, the first in point geometry, the second in line geometry. Klein's doctoral dissertation was on a particular aspect of line geometry.

His early work on geometry endeared him to this area of mathematics for life, but he did a lot of research in other branches of the subject; in particular, in the theory of functions and the study of the solutions of various kinds of differential equations.

In 1869 he made his first visit to the University of Göttingen. This university, situated in what was then a beautiful garden-like town, was to attract Klein not once but twice again, lastly as a great and pioneering force within the university and the whole of mathematics.

Following his first visit to Göttingen he went to Berlin and then to Paris, in the way that European and especially German research students were apt to do in the course of learning the new developments of the time. The stern professors at Berlin did not give Klein the kindest of receptions; indeed it was said by some of them at the instigation of Weierstrass, that what Klein did was not mathematics. In Berlin he met the Norwegian Sophus Lie, who distinguished himself with the notion of the continuous transformation groups which bear his name. The two students went on to Paris, where they lived in the same building and often worked together. Klein's attitude to geometrical research was very much influenced by the ideas that the two discussed during their time in Paris. As well as other mathematicians, they met Camille Jordan, who championed Galois' theory of groups. He published his own work, *Traite de Substitution* which explained and extended this, and brought it to the notice of the mathematical community. This book, which was to fascinate and influence both Klein and Lie, had then just been published. Its influence on Klein was to be seen in his later approach to geometry. Unfortunately, the period in France was ended by the outbreak of the Franco-Prussian war. For a short period during the hostilities Klein served as a medical orderly, but was invalided home suffering from typhoid fever.

At the beginning of 1871 he was appointed unsalaried lecturer or Privatdozent at Göttingen, having undergone a process of Habilitation by submitting an advanced thesis. During the next year after attempting to make a living from the course fees of students, the twenty-three year old Klein was sufficiently distinguished within mathematics to receive a 'call' to the position of Ordinariat, or full professor, at the University of Erlangen. Erlangen was not one of the great German, but Klein's appointment was still a meteoric promotion since he missed out the stage of being an Extraordinarious. Such things did not often happen in Germany.

After his appointment the new professor had to deliver an inaugural lecture in which he outlined his position within the subject and the direction of his research. Klein's lecture, his *Erlanger Programm*, was the most classic and distinguished of its type. It was not only given orally and published in *Mathematische Annalen* in 1873, but was translated into six languages, (English, French, Russian, Polish, Italian and Hungarian) and published throughout Europe and America. In English it was given the title of *Comparative Considerations of Recent Geometrical Researches*. In this statement he unified many of the geometries then known by expressing them within the framework of group theory. Later, he was to extend his scheme to other geometries and to show how the relativity theory of Albert Einstein, and other branches of theoretical physics could also be included. The *Erlanger Programm* was influential in stimulating the further development of geometry and its influence continues today, even though many of the new geometries of the twentieth century do not fit easily, if at all, within its framework.

The spirit of the *Programm* will be seen if we consider only a simplified version of the classification scheme. We will return to the most general of geometries later but one which is very fruitful for further development is the geometry which preserves straight lines during transformations. If we think of a film projector projecting an image of a film frame we know that, given the right conditions, we can obtain an enlarged version of the same frame. However, many other sets of conditions will give us an image; anyone who has attempted to set up the equipment for a film show will know just what peculiar images can be obtained. The set of all the possible images is obtained by a set of transformations, using the light of the projector, which form a group. This geometry, in which lines are transformed only into lines, which need neither be parallel nor the same length, is known as projective geometry.

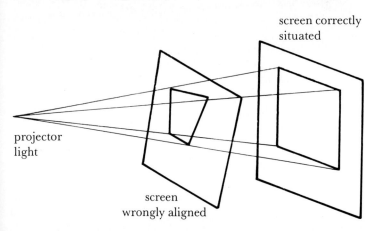

88 FOUNDERS OF MODERN MATHEMATICS

For Klein, a group of transformations possesses the property that any number of transformations can be combined into a single equivalent transformation. The property of associativity, the existence of an identity and inverse transformations are also part of the group structure.

If, in our experiments with the film projector, we concentrate on those transformations which give images where lines which are parallel on our film frame are also parallel on the screen, then we will have a subgroup of the projective transformations, and a geometry in which parallelism, but neither length nor angle, is preserved. This is known as affine geometry. However, Klein did not expressly mention this geometry in his *Programm*.

We could go on and produce a further subgroup of transformations in which similarity, parallelism and angle size is preserved. These transformations will be suitable for our film show. There will be an unlimited number of these, just as there were in the projective and affine geometries from which we would select one suitable for the size of the room.

If we altered our projector so that it produced images the same size as the film frame and, as with the geometry of similarity in the last paragraph, we kept our screen parallel to the frame, we would obtain an image in which sizes, as well as similarities, were preserved. Areas would also be preserved by these transformations, although there would be others which would also preserve areas. The geometry in which length, area and similarity are preserved is usually known as Euclidean geometry. He wrote many papers on the subject and published in 1889-90 his two volume book *Nicht-Euclidische Geometry*. This subject departs from that of the famous Greek mathematician by taking exception to his fifth postulate. A postulate is taken as a basic notion of the geometry and should, according to Aristotle, be accepted as true if the theorems of the geometry are in accord with common experience. The fifth, or parallel postulate, allows only one line to be drawn through a point A so that it is parallel to a given line.

```
                            A
- - - - - - - - - - - - - - - | - - - - - - - - - - - - -   one line

_____   given line
```

This apparently innocent notion, was the cause of consternation even among some of the later Greek mathematicians. From ancient times there were numerous attempts to either prove the postulate, or to show that it was a consequence of Euclid's other postulates. None of these were successful, but progress was made once it was realised that it might be possible to have a consistent geometry where the other postulates were true but the fifth was false. This area was explored by Lobachevsky, Janos Bolyai and Gauss, although Gauss did not publish his ideas, even to him they seemed rather far fetched. Their geometries allowed for more than one line to pass through the point parallel to the given line.

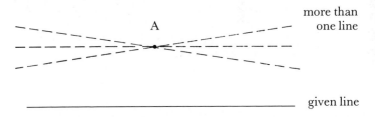

About 1854 Bernhard Riemann explored the problem from a very different point of view, and his results were equivalent to the remaining logical solution: that there was no line through A parallel to the given line.

A   no line

Klein made a number of very deep contributions to these geometries and the relationships between them, in one case extending and elaborating the research of Arthur Cayley. It was Klein who gave the names Hyperbolic, parabolic and elliptic geometry to the many, one and no line cases. However, it is rare that Euclidean geometry is referred to as a branch of parabolic geometry. Klein also showed that the elliptic geometry of Riemann could be modelled on a hemisphere, with the lines being great semi-circles which intersected each other only once.

A further unifying concept which he achieved was that the whole of projective geometry was independent of the parallel postulate, so the non-Euclidean geometries could be subsumed within this.

The most general geometry of all is topology, and topological transformations preserve only the order of the points within a line. The line itself may be distorted by the transformation.

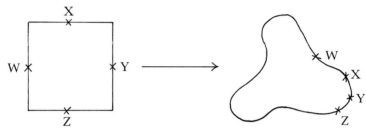

The classification of the geometries mentioned can be conveniently shown in a diagram.

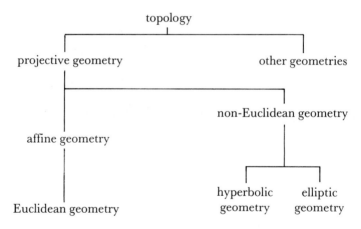

During Klein's stay at Erlangen he married the beautiful Anna Hegel who was the daughter of another professor at the University and the grand-daughter of the famous philosopher.

In 1875 he received another call, this time to the Technische Hochschule in Munich, the Bavarian capital. His friend and colleague Gordon, took on his old position at Erlangen and for many years the two travelled each Sunday to Eichstadt, a town midway between Erlangen and Munich to discuss the state of

FELIX KLEIN 91

mathematics. This was a very productive time of his life, he worked from dawn until dusk and his mind was full of ideas.

Five years later he moved to Leipzig as Professor of Geometry. During the ensuing period of his life he continued to work intensely, but he became seriously ill. At the age of 33 he was advised to take life and work very easy. His more restricted activity was to draw together many of his ideas into one of his most famous books, *Lectures on the Icosahedron, and the Solution of Equations of the Fifth Degree*, published in 1884.

The icosahedron is one of the five Platonic solids, so called because they are mentioned by the Greek philosopher Plato in his *Timaeus*. The other four solids are the tetrahedron, cube, octahedron and dodecahedron. These were the only regular polyhedra known to the Greeks. They are each made up from congruent faces of the same shape; triangles for the tetrahedron, octahedron and icosahedron; squares for the cube and pentagons

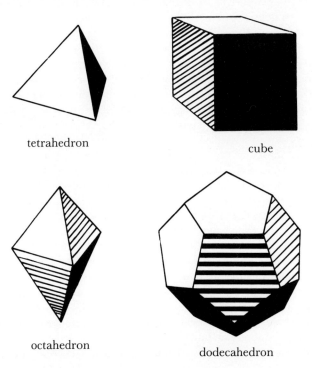

tetrahedron

cube

octahedron

dodecahedron

icosahedron

for the dodecahedron. Within any one, the angles between the faces are all equal. Although the shapes had considerable significance for the Greeks, and have been the source of interest and discoveries for many other mathematicians, there was one particular aspect of them that was special for Klein. If any of these solids is turned about particular axes, it will end up by looking as if it was in its starting position several times before the rotation is complete, but its vertices (corners) will occupy different points in space. By using an actual model and making the vertices different colours, this will be seen easily. For the tetrahedron there are twelve possible rotations, for the cube and octahedron there are twenty-four and for the dodecahedron and icosahedron there are sixty. In each case these rotations form a group. It was with the exploitation of group theory, coupled with the notion of the complex variable $z = x + iy$, where $i = \sqrt{-1}$, that Klein was able to connect the geometry of the icosahedron with aspects of the theory of differential equations, the theory of some very general functions (known as elliptic modular functions) and the quintic equation. In the first chapter of his book Klein says that the solution of the quintic equation depends on the permutation group of the relations of its roots. For the equation to be soluble, this group needs to be composed of a telescopic series of normal subgroups. The group of the rotations of the icosahedron is known as the alternating group of order five and is a subgroup of the group of the quintic equation. The whole problem of the quintic equation is also connected with the elliptic modular functions, and it was with the aid of these that Klein showed a method for its complete solution.

Remember, this is still not a solution by radicals, such a solution was shown to be impossible by Galois and Abel.

The same year as the publication of this book an invitation was received by Klein to take up the chair of mathematics in Baltimore, U.S.A. which had been vacated by Sylvester, the friend and collaborator of Arthur Cayley. This was turned down, but the next call he received in 1886 was accepted. It was to return to Göttingen, where he was to spend the rest of his life. In this idyllic town, Klein began a completely new phase of his professional life. His career as a research mathematician was to decline, but his rise as an administrator and elder statesman of the German academic establishment was to begin. In his mathematics department he organised a reading room which contained mathematical books, journals and copies of lecture notes. This new and unprecedented facility in a German university could be entered with a pass key at any time. Then, just as now, it must have been a boon to the serious student. There was also a collection of mathematical models, the construction of which had been appreciated by him throughout his career, and the collection at Göttingen is said to be extensive. As time went by he encouraged relations between the university and local industry and formed a society which had many industrialists as its members. From these connections the wherewithal was obtained to increase the university departments which had a connection with applied mathematics. His efforts with the university were so productive that he was elected in 1907 as its representative to the state parliament, the Prussian Herrenhaus, where it was felt that his horizons could be expanded even further. This, however, largely coincided with his increasing interest in secondary school education.

Klein was largely responsible, at least in the beginning, for building the mathematics department which was part of the philosophical faculty, as in most German universities, into the foremost in the land by attracting some of the greatest mathematicians of the time, and through them the most able students.

As his productivity as a working mathematician decreased, and his contact with new developments lessened, Klein began to understand how easy it was to lose touch with the new ideas and the reorganisation of the subject. To help all mathematicians to keep a grasp on the present state of their art he conceived the idea of a comprehensive encyclopaedia of mathematical knowledge, in which applied mathematics would take its place equally with the pure variety. During the years round the turn of the century, he toured Europe in search of authors with the requisite sympathy for the subject. He spent much time in England, where he believed the correct attitude to mathematics was held, and he was instrumental in having a number of English treatises on mechanics and hydrodynamics translated into German. Klein himself undertook to produce the section on mechanics. Although most of his mathematics which has been mentioned was of the pure kind, his collection of lectures on *The Mathematical Theory of the Top*, delivered at Princeton University during his visit to the U.S.A. in 1876, has become a classic of mathematical literature, albeit one which is extremely technical.

In his later years he was increasingly disillusioned by the nature of new mathematical research with its concentration on abstract structures, to the possible neglect of the applications of the subject. Klein had first begun his studies as an intending physicist and he never completely forgot this or his concern for applied mathematics as a basic tool of scientists and engineers. It was his love of applied mathematics that led him to take a major role in the founding of the Göttingen Institute for Aerodynamical and Hydrodynamical research. This was one of the first institutes of its kind in Europe and was, through Prandtl and von Mises, responsible for much of our knowledge of aerodynamics.

From about 1905 Klein became interested in the teaching of mathematics in schools, particularly in the Gymnasia secondary schools). He played a major role in making the *Meraner Lehrplanetwurfe* a plan to modernise the teaching of the subject, which included the introduction of the notion of a function and calculus into the secondary school curriculum. His influence over mathematical education penetrated further than this, even as far as England. At the 1908 International Congress of Mathematicians in Rome, Klein was elected to the chairmanship of the International Commission on Mathematical Instruction. The various national committees of this Commission were involved with either producing their own report, as in Germany, with its huge report on the teaching of mathematics in all kinds of schools, colleges and universities throughout the Germanic

States, or, as in England, with assisting and stimulating studies by their national education authorities. A comprehensive two volume Board of Education report was published in England in 1912, there were also a number of similar reports throughout the British Empire.

Klein's book *Elementary Mathematics from an Advanced Standpoint*, which was published in two volumes in 1908, was based on his lectures to German Gymnasium teachers. It examined advanced school mathematics from the point of view of an academic mathematician at the forefront of knowledge in his subject, and it acted as both a stimulus to German senior school mathematics and as a preliminary insight into the changes which were to come within the next few years. Of particular importance was that Klein advocated the use of a calculator as a teaching aid in arithmetic; admittedly this was the clanking Brunsviga machine, but the suggestion was well ahead of its time since there is still considerable debate within present education circles about what role the much more convenient pocket electronic devices are to play.

Two rather special pieces of mathematics that bear Klein's name have been appreciated by those at school and after. He was very much a master of group theory, and his use of this structure is evident in most of the branches of mathematics to which he had contributed. One group that is often mentioned at school and beyond is commonly known as the Klein Four Group. It is the symmetry group of the rectangle.

If we reflect the rectangle in the line YY′ we exchange A and B for C and D, but the rectangle appears the same. Likewise, if we reflect it in X′X the appearance is the same, except for exchanging A and D for B and C.

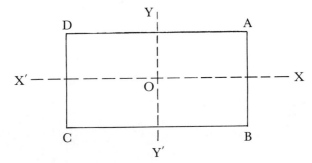

Also a rotation through 180° about 0 will exchange A with C and B with D. If we do nothing then we have the group identity. Two reflections in either X'X or YY' or two rotations of 180° will return ABCD to their original positions. Let the identity or null transformation be I, a reflection in XX' be R_1, a reflection in YY' be R_2 and a rotation of 180° about 0 be r. Then the Cayley table for the Klein Four Group will be:

	I	R_1	R_2	r
I	I	R_1	R_2	r
R_1	R_1	I	r	R_2
R_2	R_2	r	I	R_1
r	r	R_2	R_1	I

Topology has already been mentioned earlier; and one of Klein's ventures into this subject was just before his illness. Amongst other things, he described what has come to be known as the Klein bottle. This is a relation of the well known Möbius band (in which a band of material, a plane, is given a half twist and a one sided surface is produced).

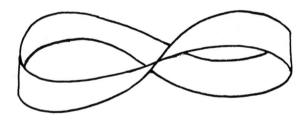

Möbius band

The Klein bottle is a tube which re-enters itself, producing a continuous one sided surface such that the bottle will not retain liquid (see photograph). Just as the Möbius band requires a three dimensional space to show off its two dimensional properties, otherwise part of it would be squashed and useless, so the Klein bottle requires four dimensions as the tube needs to re-enter itself without puncturing the surface.

During his life Klein wrote many books, only some of which have been mentioned here. Two others of more general interest were his autobiography, written two years before he died, and his history of mathematics which concentrated on developments in the subject during the nineteenth century, which he left incomplete.

Even after his retirement when he was eventually succeeded by Richard Courant, he continued to give lectures in his own house. After a protracted illness he died on the 22nd June, 1925.

Felix Klein was the archetypal German professor. He was tall and handsome; had great power within the university, considerable influence within his town and country and eminence within the academic elite of the world. He used his power quietly, in a dignified and even regal, way. Some had thought and spoken of him as "The Great Felix". Klein was not the last great mathematician at Göttingen, but he lived and worked during a period which would never recur in German academic life. He had known the humiliation of defeat in the

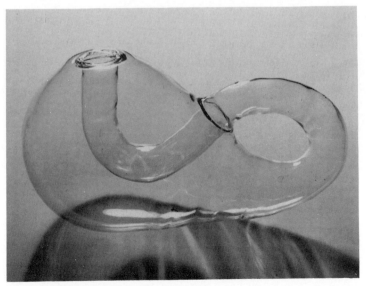

Klein bottle *Science Museum, London*

Great War but, fortunately, he never saw the discrimination in the universities, and in his country, that was to come with Hitler.

His desire for a Mathematical Institute at Göttingen, separate from and independent of the philosophical faculty, was never quite to materialise during his lifetime. Others were to succeed him and further his plans, and their own, for the university which he had developed, albeit with considerable help, into the centre of the mathematical universe.

CHAPTER EIGHT

Giuseppe Peano

AS WE HAVE SEEN, throughout the nineteenth century different sets of numbers were studied. Definitions of these were given and their properties were examined. The foundations of the real and irrational numbers were probed. New quantities such as quaternions and vectors were invented and a better description was found for the complex numbers. Using the theory of sets, cardinal and ordinal numbers were pursued to finite and infinite realms. However, even if, as Kronecker said, "God invented the natural numbers...", then no one had really come up with a popularly accepted explanation of how they work. Since the other sets of numbers are usually derived from this set, then it was important to know how it worked. The person who gave us the normally accepted version of the working of the set of natural numbers was Giuseppe Peano.

Giuseppe Peano, as his name suggests, was Italian. He was born in a farmhouse called 'Tetto Galant' about three miles from Cuneo in the northern part of the country on 27th August, 1858. His father, Bartolomeo, farmed the field attached to the house. First he attended the village school in nearby Spinetta, then, when he was older, he went to school in Cuneo, walking the three miles each way. Later the family moved temporarily to this city to lessen the burden on the five children. This must have been quite a sacrifice to make, but Bartolomeo and his wife Rosa gave every possible consideration and benefit to their family. The children gained because of this; apart from the success of Giuseppe, one boy became a successful surveyor and another a priest.

At school Giuseppe showed such ability that his mother's brother, a priest and lawyer in Turin, invited the boy to join him. He now received tuition from his uncle and at the Instituto Ferraris. First he took some examinations at the Ginnasio Cavour in 1873, and then as a pupil at the senior secondary school the Liceo Cavour, he passed the examination for the licenza liceale in 1876, which entitled him to enter the university. He passed this examination with sufficient distinction to gain a room and board scholarship at the Collegio delle Provincie. Both the scholarship and the Collegio were intended to help students from rural areas and often impoverished families to attend the university. Peano did not see either his new life with his uncle, or his academic success, as a way out from the hard organised life endured by his family; he always loved the soil and farming and spent all his holidays working on it. Also much of his spare time was spent in the country. Later in life he bought a small villa in the countryside some way from Turin.

In October 1876 Peano became a student at the University of Turin. His intention was to become an engineer and he studied design and chemistry as well as algebra and analytic and projective geometry. The following year he added descriptive geometry, calculus, physics, mineralogy and geology to his studies. At the end of that year he won a prize which exempted him from fees at the Royal Engineering School, but he decided to elect for mathematics and obtained the same financial advantage. His third and fourth years at the university were distinguished with similar academic and financial success, culminating in his graduating dottore in matematica, with high honours, in 1880.

Peano, being the best graduate in his subject that year, was given the position of university assistant to Enrico D'Ovidio. His duties were to attend the lectures of his professor and then, with the aid of his own lectures, teach the first year students, who also attended D'Ovidio's lectures. Peano was reputed to be a good teacher; he was diligent with his preparation and methodical and accurate in his lectures. The position of assistant was intended to help able scholars to begin an academic career and develop a reputation by research in their subject. Peano was to find it difficult to preserve all the time he wished for his research since, as assistant, he did much of the routine work, and then had to spend a great deal of time preparing his lectures.

The following year Peano became assistant to Angelo Genocchi and taught the second year calculus class. His earliest discoveries were in calculus and it was with these that he was to make his early mathematical reputation. Fortunately for Peano, Genocchi suffered an accident and during his protracted absence Peano was in sole charge of the calculus course, giving the principal lectures as well as the routine instruction. This state of affairs was to last for two years. In the meantime, Peano prepared his own calculus textbook *Calcolo differenziale e principii calcolo integrale* based on the lectures of Genocchi. In reality this very fine book owed much more to Peano than to his superior. In the same year as its publication 1884, Peano was made privata docenza and judged qualified to carry out his own lectures as libero decente and be promoted to a senior university position. His own course dealt with the geometrical applications of calculus, and his research was concerned with aspects of the same subject. Apart from the new results he discovered in calculus, he was a lifelong corrector of the mistakes of others, often becoming involved with long discussions in print over textbook inaccuracies or shortcomings of clarity or brevity.

In October 1886 he began a long association with the nearby Royal Military Academy, where he was appointed a professor and again taught calculus. His academic career was now moving along well, even though he was not yet a professor at the university there was every prospect that he would soon become one because of his leading role in the calculus course. With this comfortable situation he married Carola Crosio, whose father was a painter and illustrator of some note. They lived in Turin, and his new wife frequented the opera. The Peanos are known to have attended performances of Puccini's *La Bohème* and *Manon Lescaut*, but Giuseppe was obviously not that keen on opera since he is known sometimes to have slept during the performances.

At the end of 1890, after Genocchi had ceased lecturing and finally died, and the lengthy process of appointing a professor had run its course, Peano was appointed Extraordinary Professor of Infinitesimal Calculus. In the meantime, however, he had made nearly all the discoveries for which he is now remembered, although he believed that his later work was more important.

His three remarkable areas of achievement are, for our purpose, best considered out of chronological order. It is difficult, even now, to judge which of his discoveries was most significant.

In 1889 he published a pamphlet of about thirty pages entitled *Aritmetices principia, nova methodo exposita*. This title is in a somewhat peculiar version of Latin, and means *The Principles of Arithmetic by a New Method*. Peano had become increasingly interested in logic throughout his life and this booklet relied heavily on his own logical formulations, which he believed were an improvement on those of Boole and the later logicians, Pierce and Schroder. After a preface devoted to logical principles he gives a list of nine axioms, of which four are concerned with the meaning of the " = " sign. The other five axioms are now known as the "Peano Axioms", and are taken by mathematicians as a convenient starting point from which to develop the common sets of numbers. Peano's important idea was to consider a "successor" which was a way of getting from one natural number to the next. The axioms or undefined statements are:

(1) 1 is a natural number; 1 ε N, the set of natural numbers.
(2) Every natural number has a successor, a + 1.
(3) Two natural numbers a and b are equal if their successors are equal, i.e. if a + 1 = b + 1.
(4) 1 is not the successor of any natural number.
(5) If a set K of natural numbers includes 1 and if when K includes any number x it also contains its successor, then K contains all the natural numbers.

The last axiom is the principle of mathematical induction.

It will not be surprising to learn that Peano's next statement after the axioms is one of definitions in which: 2 = 1 + 1, 3 = 2 + 1, 4 = 3 + 1 and so on; and his first theorem is that "2 is a natural number", i.e. 2 ε N, which he proves from his axioms. He later modified his axioms to begin at 0 instead of 1, and thus included 0 as a natural number. However, the version using 1 is still the most popular. Peano's treatment of the natural numbers was taken up by mathematicians partly because of its simplicity and partly because it was adopted by Bertrand Russell and published in his influential logical works especially in *The Principles of Mathematics* (1903). He was not, however, the only mathematician with a basis for the natural numbers. Dedekind published what amounted to a very similar idea in 1888, but it was cloaked in a deeper and rather more obscure mathematical background. It neither appealed at the time, nor has it been taken up in any great way by later mathematicians. The logician, Gottlob Frege, also had another theory of the natural numbers. During the present century there has even been a treatment of the topic by Schmidt which used a "predecessor", the very reverse of Peano's idea. Schmidt's axioms can be derived from those of Peano.

Among the many other strange facts which Georg Cantor discovered was that published in 1878 which showed, in effect, that there were as many points in the whole of the area of a square as there were in a line of the same length as the square's side. (This strange fact also holds good between the points of a cube and those of a line the length of its edge.) As a consequence of this result, it should be possible to draw a line, a curve, through every point of the square and thus find a curve which completely fills its space. In 1890 Peano announced in print his discovery of a space filling curve. At first he gave no pictures, believing that the correctness of his proof would be disputed unless he could prove its existence using only mathematical symbols. It was left to others to produce the first sketches of such a curve.

Unlike the plotting of just one curve, it was the plotting of a whole sequence of curves, each filling more points of the square than the preceeding one. Peano was able to show how to get the new one from its predecessor, and that eventually, is the limit (in very much the same way as we use it in calculus), there is an ultimate curve in the family which passes through all the required

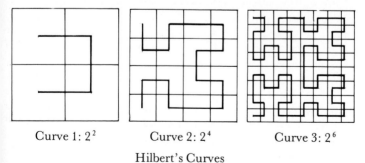

Curve 1: 2^2 Curve 2: 2^4 Curve 3: 2^6

Hilbert's Curves

points. In fact, a whole family of curves with this limiting property of filling space exists; one of the simplest to illustrate is that due to David Hilbert in 1891, about whom we will read more later.

By subdividing a square into successive even powers of two Hilbert was able to show that in the limit, when the small squares coincide with the points of the whole square, the last one in the series, the ultimate one, would go through all the points in the large square. Peano later published a drawing of his curve which amounted to a similar idea, but dividing the square into small squares in a sequence of even powers of 3. His curve 2 passed through 81 small squares.

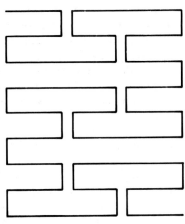

Peano's Curve 2 (81 squares) 3^4

Strange curves have been produced by other mathematicians too. Von Koch produced the well known snow-flake curve based upon the division of the sides of an equilateral triangle successively into three parts, and drawing new, smaller triangles on each middle segment. This curve has the property that the distance between any two points on the curve is infinite. Sierpinski produced a curve very like that of Hilbert which contained every interior point of a square, but, unlike Hilbert's, it was a closed curve with neither beginning nor end. Also it encloses an area less than half that of the square.

Peano published two books with "geometrical calculus" as

The snow-flake curve after two subdivisions

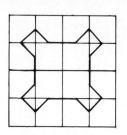

The repeating motif of the Sierpinski curve

part of the title. The first was based on his course on the geometrical applications of calculus; the second was a deeper work which involved deductive logic and contains, in one of the later chapters, the earliest statement of the axioms of a vector space. These spaces now have many applications in geometry and algebra. In Peano's geometry they originated in a form very much like that we use now in the sixth form or at the beginning of the university course.

Peano wrote of the 'entities' **a** and **b**, rather than the vectors **a** and **b**. He stated that these 'entities' must satisfy the axioms of an Abelian group, that is closure, commutativity, associativity, the existence of inverses and an identity **0**. Also he mentions real numbers m and n et cetera which are distributive with respect to addition over his 'entities' and satisfy the associative law as coefficients, and which include a unit element, i.e. $m(\mathbf{a} + \mathbf{b}) = m\mathbf{a} + m\mathbf{b}$, $(m + n)\mathbf{a} = m\mathbf{a} + n\mathbf{a}$, $m(n\mathbf{a}) = (mn)\mathbf{a}$, $1\mathbf{a} = \mathbf{a}$. He called these structures of his 'entities' with the real numbers 'linear systems'. Peano even mentions the possibility of extending these to infinite dimension systems.

Generally Peano approved of and encouraged the use of the newly developing vector methods at a time when many of his senior colleagues, and other important mathematicians, found them either troublesome or of little importance. He was a founder of the International Association for Promoting the Study of Quaternions and Allied Systems of Mathematics.

Many of the great mathematicians and scientists of the last century founded, or at least attempted to found, a journal in which learned papers on their subject could be published. The founder of a journal was, through his interests and editorial policy, able to influence the articles that were published and ultimately, through what was published, the development of his subject. Peano was no exception; on his promotion to Extraordinary Professor he set about establishing the *Rivista di Matematica*. The flavour of this journal mirrored Peano's interest in mathematical logic, especially its use as a tool for the precise expression of mathematical formulae and theorems. In fact, *Rivista* became so increasingly devoted to this purpose that Peano created a formulary of mathematical truths. The early parts of this were appended to issues of his journal, but the project soon outgrew this and a separate publication began. His *Formulario* ran to five editions, and listed the important formulae and results from most areas of mathematics, often with detailed historical notes. Peano had many collaborators who were much younger professional mathematicians from the University of Turin and the Military Academy and who contributed extensively to this large undertaking.

Peano quotes Leibniz, the philosopher and co-inventor of calculus, as one whose desire it was to see a universal means of expressing complex ideas in more simple terms according to fixed rules. Peano achieved this, but he had many things against him. His notation, although basically simple and certainly useful, was new, and it was necessary first to learn it in order to interpret the contents of the *Formulario*. Lists of formulae, no matter how complete and accurate, are very indigestable. Probably also, while Peano reached the zenith of his power as a mathematical logician with his *Formulario*, Russell went on to develop the subject further, and in a new direction, and eventually displace him as the leading exponent of mathematical logic. The first edition of the *Formulario* was published in 1895 and written in French, as were the next three editions although the title varied slightly. The final and magnificent fifth edition was published in 1908 and contained 4,200 formulae and theorems. It was, however, written in a new language of his own invention, Latino sine flexione, or Latin without grammar. Peano believed this to be his greatest creation, but, unlike his space filling curve and the axioms for the natural numbers, it is barely remembered.

Following Peano's appointment as Extraordinary Professor,

his influence and reputation grew, especially with his most famous works becoming known to the mathematical community. In 1891 he was elected to the Turin Academy of Sciences and in 1895 he was made a Knight of the Crown of Italy, the first of several honours of this kind he was to receive. This important period of recognition culminated in his promotion to Ordinario, ordinary professor, in 1895.

In 1894, a physiologist and photographer, showed a sequence of photographs to the French Academy of Sciences in which a cat that had been dropped with its back pointing upwards was shown to roll into an upright position and land on all four paws. At the Academy the event was disputed since the cat appears to contravene an important principle of mechanics, the conservation of angular momentum. Interest in the cat at least was shown by the popular press, and in all there was quite a public debate about the unfortunate, or fortunate cat. How did the cat turn over and land correctly? This was the question which Peano answered cunningly, and perhaps obviously and amusingly to the general public. It turns its tail in the opposite direction to its body until it is upright!

During the second half of the 1890's Peano was one of the world's leaders of mathematics and mathematical logic. He held offices with international bodies and played a major part in the International Congress of Philosophy in Paris which preceded that of mathematics. He was frequently president or vice-president of sections which dealt with logic. In 1900, Peano had reached the height of his career and influence in mathematics. At the Philosophy Congress he met Bertrand Russell and so impressed him with his logic and its notation that Russell did not stay for the second Congress but went back to England to study Peano's work. This meeting was to give Russell the means of examining the nature of mathematics which he used in *Principles of Mathematics* (1903) and in *Principia Mathematica* (1910). From this time Russell was to take over as the most influential mathematical logician and Peano's importance began to decline. Mathematical logic developed into a very fertile area for discoveries and it was exploited by many other people during the first few decades of the twentieth century.

At the Congress of Mathematicians in Paris, Peano refuted the suggestion made by David Hilbert in his famous address, that there was a problem as to whether the axioms of arithmetic are consistent, that with the rules of arithmetic it could always be decided whether a statement was true or false. Peano said that a paper to be presented the next day by Padoa would satisfy Hilbert that the consistency of arithmetic had already been established by his friends in Turin. However, Padoa spoke on another topic and Hilbert did not attend the session; so honour was not satisfied on either side. This is not the end of the story as we shall see in later chapters.

It is often said that we cannot just hold on to what we have, we either advance or decline. Peano began to decline as far as his mathematical reputation was concerned. Concern was expressed within the Military Academy and within the university that he was not teaching in a manner suitable for his students. He now relied entirely on his *Formulario* and his lectures were mere explanations of the precise but formal and stagnant formulae it contained. This might have been a suitable method for able and dedicated intending professional mathematicians, but most of his students at both institutes had other intentions, often in engineering. Peano's post was secure at the university; it was neither possible to dismiss a university professor from his position of great power and influence, nor was it good politics to attempt to get rid of one of Italy's most accomplished academic sons. However, things were different at the Military Academy and Peano was placed in a position where he had to resign. It is not really known how he felt about his dismissal but he was probably so busily involved with his *Formulario* and other projects that he appreciated the increase in time free from lectures. At the university the engineering school organised their own mathematics lectures to get their students away from Peano.

One of the other projects was the commission of his services by a pension fund which had run into financial and accounting difficulties. Peano, who had had no formal training in actuarial skills, produced a flawless report which convinced the Minister of Industry and Commerce that the fund should be saved. He went on to become an expert in this field and published several articles on actuarial mathematics. Also he suggested some alteration in the law for the improvement of pension funds.

Peano came from a family of small, and almost certainly, poor farmers. He loved the open air and the countryside, with its space for walking and the exercise of his numerous dogs. So that he could get away from the city and practice this kind of recreation, he bought a small villa at Cavaretto in July 1891. He always had sympathy for the workers in the fields and those who earned only a small wage by long days of back-breaking toil. It was not surprising that his politics were socialist, and he was not afraid to let his views be known. This cannot have endeared him to the general academic populace of relatively rich and powerful professors. During 1906 the cotton workers in Turin went on strike for, among other things, a normal working day of ten hours. Peano invited the strikers, who were nearly all women, to walk out to Cavaretto and enjoy the hospitality of his villa. Several thousand are said to have made the hike.

Also at his little villa he kept a printing press, which he had used at times for both his journal and the *Formulario*. Peano was concerned with the difficulties of printing logical and mathematical statements. He became an expert in the printing of mathematics and, even when the press was not used, he set up the type for some of his works at the villa.

During the later part of the last century difficulties of communication between people in different countries were caused by the lack of a common language. This is hard to understand now, when English is the leading language of science and commerce; so that even some national scientific journals from Japan and Germany are published in English because it is the first language of most of their readers. One hundred years ago there was often limited communication between various groups of people within one country. Generally, however, those who operated at an international level, either as scientists or businessmen, had a good command of the appropriate languages. Peano was no exception, he was an excellent exponent of the languages of Europe; he could write in French as well as many Frenchmen and he published articles in Spanish, Portuguese and English as well as translating the works of the German mathematicians into Italian. As an expert linguist he could both appreciate the difficulties of communication and command a means for their solution. He believed that an international auxiliary language was the answer.

There had been other international languages. Volapuk, the artificial language invented in 1879 by the priest Johann Martin

Schleyer, was already accepted on a fairly large scale. At one time there had been over a million users, and there were numerous magazines written in it. This invention was followed in 1887 by Esperanto, the creation of Lazarus Ludwick Zamenhof. (It was originally published under the authorship of ''Dr. Esperanto'', hence the name.) Zamenhof used it first as a means of communication between the Russians, Poles and Germans who lived in his home town of Bialystok in Poland. Esperanto gained in popularity over Volapuk, but it had the same drawback; it was an artificial language.

Peano was interested in an international language, but in one that already existed, not one which was artificial. He believed there were many words which were common to most of the main European languages. It was just a matter of finding which they were, and he believed that when this was done, a fairly simply universal language could be constructed. His new language, Latino sine flexione, or Latin without grammar, was announced in 1903 but he had probably been interested in this idea for a long time before. His final edition of the *Formulario Mathematico* was written in this language, as were many of his publications from then on. Most of Peano's time and energy became devoted to his new language. At the International Congress of Mathematics at Cambridge in 1912 he tried to give his lecture in it, but was prevented from doing so because it was not one of the officially recognised languages. He finally gave his address in Italian with possibly the same effect; very few of the participants, except the Italians, were sufficiently adept in it to understand his words.

In 1908 Peano offered himself as a member and as president, at the same time, for the academy that had once been devoted to Volapuk. Now it was concerned with finding a universal international language. Peano was accepted and, under his presidency, the academy went on to develop and popularise Latino sine flexione. He published a vocabulary of the Latin words which were included in the European languages in 1904. There was a second enlarged edition of this in 1909, and a magnificent third edition published in 1915 entitled *Vocabulario commune ad latino-italiano-francais-english-deutsch*, which had about 14,000 entries and was not confined to the languages of its title but also included some from Greek, Spanish, Portuguese, Russian and the Indo-European languages including Sanskrit.

This he believed to be one of his greatest accomplishments, second only to the *Formulario*. The academy was now known as Academia Pro Interlingua and it became obvious that the interlingua involved was going to bear a strong resemblance to Peano's natural language since one of the rules for its construction was that it was to use every word common to Latin, Italian, Spanish, Portuguese, English, German and Russian. Although he had great faith in his own creation, he was generally dedicated to the concept of an international language no matter who created it. He was a tireless champion of this cause and even wrote an article about Esperanto in Latino sine flexione.

The international language movement still exists, but the need for such a language is not so pressing now and the movement as a whole was dealt two enormous blows by the World Wars. Interlingua recovered from the first of these only to be almost demolished by the second.

It was natural that the established scientific community in Italy should want to aid their country during the Great War; the way they did it was to produce a set of logarithm tables. Originally Italy produced a set of tables that used no more than five figures and the more extensive tables had to be imported. This was both expensive and unpatriotic. Peano was appointed to the committee responsible for their production and wrote the preface. Later editions of these tables were probably in use in Italy until recently. (Presumably the use of electronic calculators have made them unnecessry).

Peano did not believe in examining his students. In fact he did not believe in examinations at all; neither at the university nor at school. It must seem he was the friend of every student. However, his philosophy was practical, he did not believe in torturing those who were doomed to failure in their working or professional life. Many of the students he supervised during their final year were destined to become teachers in the secondary schools and Peano developed an interest in this area. He became a member of Mathesis, a society concerned with teaching mathematics in secondary schools. Through his concern for education at this level he made a number of valuable suggestions for its improvement which are commonplace throughout the world, but were then unknown.

Peano's work at Turin University and his devotion to

Interlingua went on without many changes until 1925 when Francesca Tricomi was appointed as Ordinary Professor of Complementary Mathematics. Peano was renowned as an international expert on calculus and mathematical analysis and had been a professor of infinitessimal calculus since 1890, but his interest had turned to more elementary mathematics. It was his desire to take over the course of complementary mathematics. This he did with the agreement of Tricomi and the university authorities, but he still held the nominal title of Ordinary Professor of Infinitessimal Calculus until 1931. Even though he was still very much respected as one of Italy's leading scientists, his presence was resented in the mathematics department because his teaching was far from what it had been before his obsessive concern with the *Formulario*. It is not easy, however, to separate fact from resentment, it may be that at least some of the criticism he suffered was misplaced.

On the 19th April 1932 Peano gave his usual classes and then visited the cinema on his way home. That evening he became ill and during the next morning he died from a heart attack. He was buried with academic honours in which an oration was given in the entrance hall of the university after the funeral mass in the Cathedral.

Despite the decline of his academic powers and his move away from higher mathematics in his later years, Peano must be included among the great for his contributions to the subject. He did not indulge in the merry-go-round of academic appointments, so common with the German professors, but had stayed in Turin where he is still remembered, although not always with affection.

Apart from the three aspects of mathematics which were mentioned earlier, Peano devised much of the notation of modern mathematics which we use now. The symbols for union $\cup$ and intersection $\cap$ were due to him, as was epsilon for set inclusion, but this has now become $\in$. Peano used $\supset$, a reversed C, to show a subset, where we now use Russell's $\subset$. Russell adopted much of Peano's notation in *Principia Mathematica*, and it is from this source that it has become popular with us. The symbols $\forall$ for 'all' and $\exists$ for 'there exists' are also due to Peano. His symbols ⓘ and $\bigcirc$ for the equivalent of the universal and empty sets respectively have not, however, survived. Peano is also responsible for making the simple, but important distinction between the element x and the set containing it $\{x\}$.

David Hilbert (1862-1943) *Bildarchiv Preussischer Kulturbesitz*

CHAPTER NINE

David Hilbert

DAVID HILBERT, the second great mathematician from Göttingen, achieved even greater fame than Felix Klein. His research spread over many branches of mathematics and theoretical physics and has made a lasting mark on both subjects.

Hilbert was born in Wehlan near Königsberg, the capital of East Prussia on the 23rd January, 1862. At that time Otto, his father was a county judge. Shortly after his birth his father was appointed a judge in Königsberg itself and the family moved there.

Königsberg was the birthplace of the great philosopher Immanuel Kant and many times during his childhood his mother, Maria, took him to visit Kant's grave on the great man's birthday, April 22nd. Also Königsberg was famous for its seven bridges and the ancient problem as to whether a person could, during a walk, cross each bridge just once in a circuitous journey and finish at his starting point. The problem was finally put to rest by the great Euler who showed that this could not be done, but at the same time he began the whole area of mathematics now known as topology.

David's first lessons were given to him by his mother and it was only when he was eight that he went to the Vorschule of the private Royal Friedrickskolleg. In the autumn of 1872 he entered the Gymnasium proper. This had been Kant's school and it followed largely a classical curriculum. At about this time, unknown to Hilbert, Herman Minkowski, who was to become his great friend, and another future friend, Arnold Sommerfeld, were both attending the nearby Aldstradt Gymnasium.

David's abilities were sometimes thought strange, he could explain mathematics problems that even his teachers had difficulties with, but his mother often did his school essays. However, the Friedrickskolleg was not right for him, it did not allow his interest and abilities to flourish fully and in his last school year he transferred to the state Wilhelm Gymnasium. This school placed great emphasis on mathematics, particularly geometry. His teachers recognised and encouraged his talent and he showed the promise of a great career in mathematics.

On leaving the Gymnasium he went to the University of Königsberg. The Jacobi mentioned in Galois letter to Auguste Chevalier (see p.10) had once been professor of mathematics there. Hilbert's father had wished him to read law, as he had done, but David enrolled in mathematics. During his first semester Hilbert followed courses in calculus and the algebra course in determinants. He spent his second semester at Heidelberg where he attended the lectures of Lazarus Fuchs, the renowned expert on differential equations. These equations, as we read in the chapter on Boole, involve the differential coefficients of calculus and are often very complicated to solve. Next he progressed to Berlin and the lectures from Weierstrass, Kummer, Helmholtz and Kronecker. The following semester he returned to his home university and made his first acquaintance with the theory of invariants, an area in which he was to achieve fame, from Heinrich Weber.

Herman Minkowski, who was only seventeen and had completed his Gymnasium course in five and a half years instead of the usual eight, returned from the University of Berlin where he had spent the last three semesters. Minkowski had been busying himself with the problem of representing a number as the sum of five squares, which the Paris Academy of Sciences had posed for the Grand Prix des Sciences Mathematiques. In fact he won the prize and Hilbert, although criticised by his father for doing so, deliberately sought friendship with the shy, gifted mathematician.

At about the same time Adolph Hurwitz came to Göttingen as an Extraordinarius or assistant professor. He was twenty-five years old and a very precocious mathematician. Each day

Hilbert, Minkowski and Hurwitz met at five o'clock and walked to a particular apple tree. During these walks the only topic of conversation was mathematics.

After the required eight semesters had passed, Lindermann, the first to show that π was transcendental, or could not be the solution of any equation that could be expressed in whole number coefficients, suggested that Hilbert might work on the theory of algebraic invariants for his doctoral dissertation.

The theory of algebraic invariants is concerned with the means of transforming geometrical figures using algebra so that their shape is kept the same. As we have seen, Arthur Cayley and his friend Joseph Sylvester were pioneers in this field. Hilbert completed his work, and on 11th December 1884 passed the oral examination. Even with this qualification it was necessary for a prospective university teacher to go through a further process known as Habilitation which was achieved by further research and a thesis, to become a Privatdozent or licensed but unsalaried lecturer. In order that he might make a living, whatever the extent of his academic aspirations, he took the state examinations in 1885 that would allow him to teach in a Gymnasium.

Part of Hilbert's further study was to undertake a trip. He went first to Leipzig where he attended Klein's lectures, then on to Paris with his companion Edward Study where he met Henri Poincaré, Camille Jordan and Charles Hermite. On his way back he paid a further visit to Klein, this time at Göttingen, the university where he was later to spend most of his life. On returning to Königsberg he prepared a paper on invariants for his Habilitation; the final part of the process was the delivery of a lecture on an area selected by the University from his own selection of topics. In July 1886 the training was over and he was appointed Privatdozent. There were ten other dozents, each attempting to make ends meet from the private fees of students, so the living earned was meagre indeed. Of the three courses Hilbert planned, only the one on invariants attracted sufficient students for it to continue.

At this time Hilbert attacked what had become known as Gordon's problem. Paul Gordon was then the great expert on invariants, and he was, to mathematicians, known as the 'king of invariants'. The problem was to prove the existence, or otherwise, of a particular set of invariants from which all the others could be constructed. Gordon had produced a rather clumsy proof for some special cases, but Hilbert showed its truth for all cases in so elegant a way that Gordon exclaimed: "This is not mathematics it is theology!" Hilbert sent Cayley a copy of his work which at first he did not believe, but he was quickly convinced of its truth and brilliance.

Hilbert, who was reputed to be a rather dashing young man and a snappy dresser, was frequently in the company of Kathe Jerosch. The romance blossomed and in October 1892 they were married. The following August Franz, their first child, was born.

While this was happening Hilbert was appointed as Extraordinarius. Later Lindemann left to take up a new post in Berlin and Hilbert was chosen, on the advice of Felix Klein, from a list of distinguished mathematicians to fill the vacant position as Ordinariat or professor of mathematics.

The daily mathematical walks had ceased when Hurwitz had left but they began again when Minowski returned to Königsberg as Extraordinarius. He and Hilbert were asked by the German Mathematical Society to write a report on the theory of numbers.

In March 1895 Hilbert was appointed professor of mathematics at Göttingen, again this was with the help of Klein whom he joined at the university where he was to spend the rest of his working life. His lectures were considered by the many German and foreign students (many of whom were American) to be deep and beautiful but he was often trapped by details. After regular Wednesday morning seminars the new professor and his students lunched at a popular restaurant. There was always a mathematical discussion and many of the advanced students and dozents who were in awe of Klein, became friends with Hilbert. The cleverest and most interesting of these accompanied him on long walks where, as always, the topic of conversation was mathematics.

Following this triumph Hilbert turned his attention to geometry. Although his interest in this subject was new, his basic idea had been formed much earlier. He believed that all the parts of geometry such as points, lines and planes should be described not in terms of what they are but in terms of how they behaved. Their inter-relationships were the most important part of the subject. Indeed he is reputed to have said that "One must be able to say always that instead of points, lines and planes — tables, chairs and beer mugs". Names and diagrams were not as important as the relationships between his 'objects'. The 'objects' were lines and points and the relationships were axioms or fixed rules for their behaviour. From his axioms he treated the whole of the geometry of Euclid, and the lectures he gave on this topic were published as *The Foundations of Geometry*. This book is fairly simple and has been a mathematical best seller for many years. It has been read by teachers and students, and it has so penetrated mathematical studies that its influence can be seen in school geometry today.

He also extended his work to analytical geometry and showed that if a contradiction could be produced in geometry (that something could be shown to be both true and false at the same time) then there would be a contradiction in the arithmetic of real numbers. Since arithmetic was thought to be free of contradictions he believed that both Euclidean geometry and the non-Euclidean geometries (see the chapter on Klein) were also free from contradictions, or consistent.

Hilbert's success and growing stature as a mathematician was noticed to the extent that he was asked to give a major address at the International Congress of Mathematicians in Paris in the summer of 1900. The new century was just beginning and he decided to suggest directions for mathematical research and problems to which mathematicians should direct their efforts. He took so much time preparing this address that the Congress programme was printed without it being included. When it was eventually finished it had the unpretentious title of *Mathematical Problems*. It would be very difficult to underestimate its effect on the development of the subject during the present century. The opening section is a general discussion of the importance of problems and the ways in which they have led to fruitful areas of research. Then there is a list of twenty-three problems which were unsolved. Some of their solutions have now been found to the everlasting fame of their solvers. Most of the problems are very technical and require a deep knowledge of mathematics even for their comprehension, but those numbered 1, 2, 3, 6, 13 and 18 are understandable as problems even if they are not amenable to ready solution.

The first, which we have already encountered, is Cantor's

problem of the cardinal number of the continuum. Hilbert hoped that it would be possible to show that all infinite sets had either the cardinal number of the natural numbers $\aleph_0$ or that of the continuum c. It would then follow that c is equal to Cantor's $\aleph_1$ and the continuum hypothesis would be proved. The second was to discover whether the basic rules or axioms of arithmetic were compatible. Were the axioms of arithmetic really consistent as had previously been assumed, or could it be shown that certain arithmetical statements can be both true and false? The solution of this problem was subsequently to have a shattering effect on mathematics. The third problem is easily stated "Are two tetrahedra with equal bases and equal altitudes of equal volume?". This could be shown to be true for many special cases but in 1900 no general proof had been demonstrated. However, it was the first of the twenty-three problems to be solved, with the proof demonstrated in 1901 by Max Dehn, one of Hilbert's own students.

The sixth problem was in physics, a subject which greatly interested Hilbert and in which he was later to work with considerable skill and success. The task was to axiomatise physics and thus give it the same systematic basis as arithmetic or geometry. This has proved very difficult to achieve.

In the previous chapters we have seen the various methods which have been invoked to solve the equations of the fifth degree. Explicit solutions can be obtained using elliptic functions, which are expressions similar in some ways (although much more complicated) to the familiar trigonometrical functions. Also a branch of mathematics known as nomography which invokes the graphs of families of curves to solve problems could be made to solve equations of degree as high as the sixth (x^6). The thirteenth problem was to discover the conditions necessary for the solution of the seventh degree equation (x^7).

We can fill a plane with polygons so that there is no part uncovered, and we can fill space with suitable polyhedra. Also it has long been possible to have a geometry of more than three dimensions by simply, mathematically adding more dimensions at right angles to the three we can visualise. Hilbert's eighteenth problem was to explore ways in which three and higher dimensional space can be filled with congruent polyhedra.

The whole challenge of the address, which was accepted so

readily is summed up in its very last phrase ". . . may the new century bring it (mathematics) gifted masters and many zealous and enthusiastic disciples".

When the address was finally delivered at the Sorbonne the list was shortened to ten problems and the general remarks since it was too long for the time allotted. When Hilbert had concluded his speech Peano, as we have read, took exception over the question of the consistency of the axioms of arithmetic since he believed that his compatriots had developed a suitable compatible system. There was also a system published by Hilbert in 1899. The final drama in the story was yet to come.

Hilbert was now recognised as one of the leading mathematicians of the time and was honoured by the German government with the title Geheimrat, an award similar to a knighthood. He was also given a 'call' to take up the more senior post as professor of mathematics in Berlin. After much deliberation however, he decided to stay with the very gifted research students that he and Klein had attracted to Göttingen, and at the same time he persuaded the government authorities to establish a new professorship for his friend Minkowski. When it became known that Hilbert was going to stay and Minkowski was going to arrive, the members of the Mathematics Club organised a formal drinking and smoking party. Such a party, or Festkommers, was a means of expressing very high appreciation for a professor.

Hilbert was now deeply involved in the field of integral equations. These are equations where the dependent variable appears under the integration sign in an equation. In solving or inverting an integral equation it is rearranged so that the integration can be performed. This process is by no means easy and a vast study of the methods used and the complications which arise was undertaken. There are several types of equations, one of those investigated by Hilbert was

$$f(s) = \phi(s) - \lambda \int_0^1 K(s,t)\phi(t)dt$$

$K(s,t)$ is known as the kernel and the values of λ for which solutions of the equation are known, are called eigenvalues. He was particularly involved with the theory of eigenvalues and the solutions, known as eigenfunctions, associated with them.

Hilbert's development of the theory of integral equations was particularly important because they were of considerable assistance in many branches of physics. Arising out of this and his earlier work on quadratic forms he devised a means of expressing functions in infinitely many variables. Later other mathematicians expressed this as a geometry of infinitely many dimensions which became known as Hilbert space and proved a very satisfactory tool for exploring the new developing subject of quantum physics.

Over the years the already famous, and those who were later to achieve fame, came to the mathematics department at Göttingen. These included Herman Weyl, Max Born and Carl Runge as the first professor of applied mathematics in the whole of Germany. Born, who later won the Nobel Prize for physics, was Hilberts's unpaid assistant and played an active part in the preparation and execution of his lectures, which were now often attended by several hundred students, many of whom had travelled from foreign parts to study at Göttingen. These lectures were reputed to be the best in the university but even with Born and Minkowski helping in their preparation they would still occasionally end in confusion. Hilbert was frequently able to draw deep insight and considerable development from small beginnings. He believed that the whole of the theory of differential equations could be gleaned from a study of the equations: $y'' = 0, y'' + y = 0$.

In 1905 the Hungarian Academy of Science was to award a prize of 10,000 gold crowns to the mathematician who had contributed most to the progress of the subject during the previous twenty-five years. This was named the Bolyai Prize, in honour of Johan Bolyai, and it was generally agreed that it was destined for either Henri Poincaré or David Hilbert. It was awarded to Poincaré because at that time his work extended over a longer period of time and a greater range of areas than that of Hilbert. However, the services of Hilbert to mathematics received a unanimous commendation from the awarding committee, and in 1910 he received the second Bolyai Prize.

Hilbert's next great success was the proof of Waring's conjecture: that every number can be expressed by the sum of four squares, nine cubes, nineteen fourth powers "and so on". Before he could demonstrate the proof of the theorem to the regular university seminar his friend Herman Minkowski died and Hilbert dedicated his paper to him.

As well as his interest in mathematics, Hilbert developed a passion for physics. He claimed that "Physics is much too difficult for physicists" and tried to show how close it was to his conception of mathematics by developing sets of rules and relationships for its structure. In accord with his sixth Paris problem, he was determined to axiomatise the whole of physics, but this proved too difficult. The expansion of knowledge kept this from his grasp, but there were some areas where he achieved limited success. John von Neumann who visited Göttingen and was influenced by Hilbert's work produced an elegant axiomatisation of quantum theory twenty years later. Einstein and Hilbert simultaneously solved a particular problem on gravitation within the general theory of relativity and published their results within days of each other. A friendly correspondence developed between them from then onwards.

The Great War had its effect on Göttingen. Courant and many other mathematicians who had come to Göttingen first as students and then had served as Hilbert's assistants served in the war. The realisation of Klein's plans to build a new mathematics department, although well advanced, did not come about. Despite the privations caused by lack of staff, students and resource, the most devastating effect was that the international communication necessary for advanced scholarship ceased.

As the hostilities passed, mathematics, which had become an insular and inward looking study, was about to take off in a new direction. The famous *Principia Mathematica* by Bertrand Russell and Alfred North Whitehead had been published during the years immediately before the war and the interest it generated had been strangled, but from 1919 there was considerable preoccupation with the foundations of mathematics. Naturally Hilbert involved himself and his assistant, now Paul Bernays, with research in this area. Already distinguished in this field was Jan Brouwer and he and Weyl were proponents of the "Institutionist School" which was attacking the foundations of the subject without putting anything in their place. Hilbert was a proponent of what had become known as the "Formalist School" and believed that much of the structure of mathematics could be saved if it was expressed in a system with the theorems and their proofs written in the language of symbolic logic as sentences with a formal logical structure.

Courant had returned to Göttingen and become Extraordinarius and then Ordinariat as Klein's successor. He was well known for his deviousness and ability to manipulate events in his favour. One event with far reaching consequences was his application to the Minister of Culture, who was responsible for the administration of the German universities, to alter the heading of the department's stationery from "Universität Göttingen" to "Mathematisches Institut der Universität Göttingen". He received permission for this apparently small change and proceeded to develop the department into a new, large mathematical institute, soliciting finance and putting up a new building in the process.

In 1924 Courant published the first volume of a book which was destined to influence applied mathematics for more than fifty years, indeed up to the present. The *Methods of Mathematical Physics* by Richard Courant and David Hilbert was a collection of Hilbert's work, research and lectures on his mathematical development of theoretical physics. It is Hilbert's work on integral equations, Hilbert space and the further development of more traditional methods that will be his greatest contribution to physics. Although the words were written by Courant and his students the book is popularly known as the Courant-Hilbert because it is the expression of the mathematical spirit of Hilbert put into Courant's words. The book was a considerable improvement on textbooks available at the time and is now still in print as the leading work on mathematical methods. At the time it was particularly relevant and helpful to the development of the quantum theory. Integral equations and the theory of infinitely many variables or Hilbert space were useful tools in this area.

At about the same time the Hungarian mathematician, John von Neumann, was a frequent visitor to the Hilbert household. Hilbert was always slow to understand anything new, and new ideas often had to be repeatedly explained in great detail before he fully comprehended them. Von Neumann was the exact opposite, his mind was lightning fast. In 1924 he was twenty-one and a very formidable mathematician. He was interested in Hilbert's physics, which he extended to fit the development of the quantum theory more precisely, and his ideas of mathematical proof. Hilbert believed that every problem was soluble, a solution could be found or a theorem proved. Hilbert, von Neumann and Bernays, who was now a Privatdozent but still Hilbert's assistant spent much time together, often concerned with the foundations of mathematics and the idea of proof. Bernays prepared and often delivered part of Hilbert's lectures and looked after the research students. Although the two of them often had violent arguments, usually about politics, they eventually produced the book *Grundlagen der Mathematik* (The Foundations of Mathematics). Hilbert's development of the foundations of mathematics was vehemently criticised by Jan Brouwer the leader of the "Institutionist School' who was antagonised by Hilbert, his ideas and his supporters to the degree that he would leave any company that treated Hilbert's ideas favourably. The feeling was reciprocated to a lesser degree by Hilbert.

In 1928 the Germans were invited to attend the International Mathematical Congress at Bologna, this was the first time they had been allowed to do so since the Great War. Hilbert led the delegation and they were welcomed with cheers; the world of mathematics was again united.

In 1930 Hilbert was sixty-eight years old, the age at which a German professor was required to retire and become an Emeritus professor. An Emeritus professor has the title and privileges without the duties and obligations of a member of the university staff. The German government gave such retired professors a generous pension. During the winter of 1929-30 Hilbert delivered his *Farewell to Teaching*. He returned to the topic of invariants, an area which had helped him achieve fame. At the course of his lectures there were many professors as well as crowds of students. As a mark of respect a new street in Göttingen was named Hilbert Strasse and Königsberg made him an honorary citizen.

After retiring he continued to lecture, but otherwise led a fairly quiet life except for gatherings of the family and friends at his and Kathe's birthday parties. The greatest celebration of all was to be his seventieth birthday in 1932. His mathematical works had been collected and edited, and were to be published by his friend Julius Springer. At the Mathematical Institute there was a great party with many famous friends from mathematics and physics present, many coming from foreign parts. The climax of the evening was a torchlight procession of students who came through the snow to the Institute where they shouted for Hilbert. This was the highest honour that students, some old and some new, could bestow on their professor.

Hilbert's life as a mathematician had touched upon many fields and in all he has left his mark. He moved from one area to another and rarely returned. His early work on algebraic invariants and the theory of numbers gave way to geometry, then to integral equations followed by physics and finally to the foundations of mathematics and mathematical logic. His spirit is felt today, and will continue to be felt, in the challenge he gave mathematicians with his Paris problems. Much of his influence on school mathematics and the less specialised areas of university mathematics has been indirect. The Hilbert space that bears his name was a systematising by others of his ideas in this area. Like Peano he contributed towards the structures now known as vector spaces. Other developments, like his construction of natural numbers were overshadowed by the ideas of others. In this case, Peano's ideas were so powerful that Hilbert's were ignored for many years, the paper describing them was even left out of his collected works.

In 1933, Hitler came to power and the universities were ordered to dismiss every full blooded Jew who held a teaching position. This included many leading members of the Institute: Courant, Born, Bernays, Weyl and Emmy Noether. Hilbert was left almost without mathematical friends, only Paul Bernays remained privately, at his own expense, as his assistant. Their book *The Foundations of Mathematics* was almost ready for publication. The leadership of the Institute changed several times since Courant had been required to leave; for part of the time a Nazi functionary was in charge.

Hilbert quietly lived out the rest of his life in Göttingen. As time passed his health deteriorated and he died on the 14th February 1943 having survived, at least in Germany, many of his friends and former colleagues.

CHAPTER TEN

"Nicholas Bourbaki"

AROUND THE TURN of the nineteenth century there were at least three great statements about mathematics. Klein's was the *Encyclopaedie der mathematischen Wissenschaften mit Einschluss ihrer Anwendungen* which encompassed most aspects of pure and applied mathematics, including physics, geophysics and astronomy. It was, like most encyclopaedias, well out of date by the time it was completed in 1935. Peano's *Formulario Mathematico*, which is hardly known today, was a heavy catalogue of formulae which were written in a highly condensed form and obscured by an artificial language. These two works listed what was known near the end of the last century. On the other hand, Hilbert referred in his problems to the unknown and thereby gave an impetus to mathematical progress in the present century.

The next major description of mathematics was to take a still different form and was very much a product of the twentieth century. The author was "Nicholas Bourbaki". He first showed himself on the mathematical scene by publishing notes in the *Comptes Rendus*, the journal of the French Academy of Sciences, during the middle of the 1930's. However, his name sounded Greek and his background was unknown. Later, beginning in 1939, an extensive multi-volume mathematics textbook began to appear under his authorship.

In universities and colleges in France, as in many other countries, it was common for senior students to play tricks on their first year colleagues by impersonating professors and giving misleading information or even bogus lectures. This was the case at the Ecole Normale Supérieure, one of the great training grounds for French mathematicians. An apparently distinguished but in reality bogus, visitor called Nicholas Bourbaki gave very cleverly constructed lectures of mathematical nonsense intended to deceive freshmen. At least this is how one story goes.

There was an officer called General Charles Denis Sonter Bourbaki who achieved fame during the Franco-Prussian war. His distinguished career came to an end in 1871 when he, with his dishevelled army, took refuge in Switzerland, only to be interred. With this prospect before him he attempted to shoot himself, but missed and lived to a ripe old age. He must have achieved greatness earlier, however, since he was offered the throne of Greece in 1862. Is this significant? The ancient Greek mathematicians have certainly been venerated and in the case of Pythagoras, almost deified. There was said to be a statue of the General in Nancy, and certainly at least some of his modern namesakes are known to have spent time at the University of Nancy.

Another version of the story, is that "Nicholas" and his book originated in a conversation between the French mathematicians Andre Weill and Jean Delsarte about the most effective way to teach calculus. This story is reminiscent of the one which describes Sir Isaac Newton discovering the universal law of gravitation because an apple fell on his head. It may be the truth, but nothing like the whole truth.

The real story of Bourbaki starts with the Great War. Many of the scientists and mathematicians who achieved fame during this century took part in the conflict which began in 1914. Courant and von Neumann, who were later to spend time at Göttingen, served with the Germany army. Littlewood, the famous twentieth century English mathematician, spent his time as a lieutenant specialising in ballistics and calculating the trajectories of shells. Most armies attempted to employ mathematicians and scientists where their talents would be useful. This was not the case with the French, and many promising mathematicians perished in the hostilities. Those who were too young to take part, like Weill, Desarte, Jean Dieudonne and Claude Chevalley, became conscious of the loss of a generation of French academics. The older professors who were not involved with the fighting, taught only the mathematics which was important to them. Important mathematical areas which had been developed only one, two or three decades earlier did not form part of either undergraduate or postgraduate studies, and it was a revelation to the new generation of French mathematicians when they heard of these from other sources. The French had boasted amongst their number men like Poincaré, who had been adept at all branches of the subject and now it seemed as though the greatness and general all-encompassing character of French mathematics was coming to an end. To save this from happening and to fill the gap left by the destruction of war, "Nicholas Bourbaki" was created. His main object was to be the production of a book in French which systematically described those parts of mathematics which were seen as important and had been neglected in France because of earlier unfortunate circumstances.

There was to be a unification and simplification of mathematics. Very often progress in mathematics, or any of the sciences, proceeds when an idea unifies a whole collection of more diverse and disparate facts into one simple but powerful statement. Bourbaki was to do this in three years, according to the original objective. However, the enterprise has been going for more than forty years and is not complete yet.

The *Elements de Mathematique* was to include all those results which a working mathematician would require. It was to be an advanced textbook suitable for students, and it was to be a simple unified reporting of important facts. To achieve this Bourbaki took drastic action. It was harshly selective in what it considered to be the kernel of mathematics; and it was very abstract in what it considered to be mathematical structures.

We have seen various mathematical structures develop during the nineteenth century, particularly the group, but also the ring and the field. Bourbaki has ignored most of these and describes in their place only 'algebraic structures' in which the rules of composition are the important relations. There are also order structures, in which the order relation is important as with the real numbers, and topological structures, where notions such as limit, continuity and neighbourhood play an important role. Each structure is made up from sets of elements, relations into which these elements enter and axioms. Axioms are conditions which are defined to hold between the relations.

Each structure has its own collection of theorems. These are very general and very powerful since they can be applied to diverse branches of mathematics which have the same underlying structure. We have seen in the chapter on Hamilton that complex numbers can be used in Euclidean geometry. The theorems of both these topics can be seen as equivalent and, in Bourbaki's terms, they are examples of those pertaining to a more general structure to which Euclidean geometry and the algebra of complex numbers both belong.

Frequently two or more of the structures can be combined to give a new structure which forms the basis of a different area of mathematics. For example, the branches of calculus emerge from a combination of the algebraic structures with those of order and topology.

In terms of the way we commonly think of mathematics, Bourbaki's *Elements* is a very peculiar book. The contents are very much a product of this century, yet the authors concentrate on fundamental established mathematics, recasting it to fulfil their objectives. It is mathematics picked clear of all applications, even those within mathematics itself; it is mathematics with its skeleton laid bare.

As we might expect the traditional divisions of the subject have been eroded. There is a part entitled *Algebra* and one of *Set Theory*, this we might expect from previous developments, *Functions of a Real Variable* and *Topological Vector Spaces* are more modern areas.

All this may encourage us to think that this is a very dull and difficult book. However, it is by no means as dull as it sounds and Bourbaki has made every attempt to make it a book suitable for study. There are numerous exercises, some of which include old problems recast in Bourbaki's mathematics. Historical notes lighten the text and place ideas in perspective. A sign ζ is used to indicate those parts of the text where there is a sudden switch in the argument or where generally greater care is needed to keep a grip on the subject. New and simplified terms were introduced, many of which went against current usage. Also the foundations of the subject, which so troubled Hilbert and Brouwer, have largely been ignored.

Bourbaki is not without its critics, mainly because of the rarified and highly abstract nature of the *Elements*, but its influence and success during the last few decades has been enormous. This is no doubt due in part, to the great care which goes into its production.

"Nicholas Bourbaki" is a group of about 10-20 mathematicians. Except for Samuel Eilenberg, the Polish American algebraic topologist, they have been exclusively French. Two or three times a year the group meets in pleasant surroundings with good food and wine, which are paid for out of the ample royalties from the *Elements*. At each meeting, work in hand is discussed and new additions are allocated to authors. Descriptions of these meetings indicate an almost riotous level of activity. Drafts of work to be included in the *Elements* are mercilessly pulled to pieces. They are read aloud and everyone is free to join in the discussion, sometimes three and four at once. There is little respect for age or seniority, and a good command of the French language is important, which is one reason why the group is almost exclusively French (Eilenberg speaks the language like a Frenchman). Any of the group can prevent a draft from going for publication and it is not infrequent for ten or so drafts to be written. The authors must give way with good grace to the veto of any member. Usually another author is appointed, then another and so on until no more progress can be made. The period of time between inception and publication of each part is usually twelve to thirteen years.

There have now been several editions of some parts of the *Elements*. These subsequent editions have usually been produced far quicker than the original. They have attempted to keep pace with changes within the subject as the book is intended to represent the best way of presenting it in its present state, even though the mathematics is only the underlying areas in which expansion has taken place and not the areas of research themselves. To produce these editions it is important to pursue the ideas of those mathematicians who are at present active in developing and interpreting the subject; that the book should be ahead of those whose horizons have become fixed is one of the reasons for its production. For this reason the one rule of Bourbaki is that members retire at 50. The founders have now gone but the *Elements* still flourishes.

New members are admitted to replace the retiring ones, this ensures that "Nicholas Bourbaki" continues and that the *Elements* is a work of its time. New members are found amongst the very able research students and junior academics by the established members, who are usually their senior colleagues in the universities. A member has to have two qualities: he must speak and make his views known, the silent partner does not exist, and he must be prepared to work in all branches of pure mathematics both critically in the discussion and actively in writing, possibly about something in which he is neither expert nor particularly interested. In short, he must believe in the unity of the subject and be prepared to work hard to achieve it. Perhaps a third, more personal trait, is that he must be able to withstand criticism well.

"Nicholas Bourbaki" has published other things, usually papers describing its approach to mathematics. One, *The Architecture of Mathematics* (published in 1950 in the *American Mathematical Monthly*) has a footnote describing "Professor N. Bourbaki, formerly of the Royal Poldovian Academy, now residing in Nancy". Another gives his address as the University of Nancago, Nancy plus Chicago (Andre Weill was at the University of Chicago). It is believed that "N. Bourbaki" attempted to gain membership of the American Mathematical Society. The important learned body turned down the application with the facetious suggestion that it could apply for institutional membership.

A body like "Nicholas Bourbaki" is in a special position when it comes to its own existence. It can and has played tricks, of which the one just described is an example, and perhaps it has invoked the wrath of its critics. The one great advantage it has is possession of a continuing set of objectives. That these objectives are important is seen in the survival and prosperity of the single name which unites them and takes the place of more than a dozen changing names on the title page of the *Elements*.

The influence of "Bourbaki" has been great. Few mathematicians will have completed their training without some contact with its work even if only at second hand from their university teachers. They will then pass these ideas on to their students, whether at university or school. Not everyone will like or accept "Bourbaki's" philosophy but twentieth century mathematical thinking cannot ignore it. The filtering of its ideas through to school level has been felt throughout the world. There has been a tendency towards abstraction at all levels of mathematical education and although "Bourbaki" is by no means the only cause, he has contributed towards it.

Postscript

THE DISCOVERIES AND CREATIONS of the mathematicians we have mentioned were taken up and experimented with by others. Some things, like matrices and vectors, fared well and went from strength to strength. With others there were pitfalls and surprises.

The set theory of Cantor was one of the earliest casualties. Even when Cantor was developing the subject, paradoxes in the theory of sets were becoming apparent. One of these was discovered in 1902 by Russell, who was busy exploring the foundations of mathematics when he came across a contradiction. He examined the set of all sets that are not members of themselves. Is it a member of itself? If it is a member of itself, then it is not. If it is not a member of itself, then it is. This contradiction dealt one of a series of blows to set theory. The theory was not as simple as it first appeared and many logicians and mathematicians have worked on its description during the present century. However, it has survived and, taken at a simple level, it has proved a firm basis for the mathematics taught in many infant, junior and secondary schools.

Arithmetic, and the whole of mathematics, seems to work. We normally expect to be able to find the solution to a problem, provided it is amenable to mathematical methods, and to have confidence in it. We expect also to be able to ascertain whether or not an arithmetical statement is true. This would seem to be in the nature of mathematics. Hilbert and Peano both believed they had come at least some of the way to showing that arithmetic was consistent. It was thus expected that a basic set of axioms and theorems would eventually be produced for the whole of arithmetic and many other branches of mathematics. Somewhat surprisingly this belief was suddenly shattered when in 1931 Kurt Gödel, a young mathematician then at the University of Vienna, proved the opposite and brought speculation on Hilbert's second problem to an end. In a very elegant argument that used both Russell's logic from the *Principia Mathematica* and Peano's postulates for the natural numbers he showed that it was not possible to decide whether some arithmetical statements were true or false within the structure of arithmetic. To decide this would require concepts and theorems from outside the logical system of arithmetic. Arithmetic and hence mathematics was therefore incomplete. Gödel's incompleteness theorem landed a further severe blow to the foundations of mathematics. Gödel left Vienna shortly after this and took up residence at the Princeton Institute for Advanced Study. He made many other important contributions to mathematical logic and was also one of the few mathematicians to effect a solution to one of the field equations of Einstein's general theory of relativity. Only recently, on 14th January 1978, Kurt Gödel died at Princeton.

Even after these catastrophes mathematics did not seem to be any the less effective at solving problems. From the practical point of view all was well, but the foundations were insecure and various attempts were made to skirt the shortcoming which Godel had shown to exist.

The incompleteness theorem was also to influence thinking on the continuum hypothesis of Cantor. That the cardinal number $c = 2^{\aleph_0}$ of the continuum was the same as the first nondenumerable transfinite cardinal number $\aleph_1$. Gödel showed in 1936 that the continuum hypothesis would not give rise to paradoxes when it was combined with a consistent system of mathematical rules. The incompleteness theorem strongly suggested that within one of the major rigorous systems of set theory it would not be possible to establish the truth of the continuum hypothesis. Paul Cohen, an American mathematician, established this conjecture as a fact in 1963. He also pointed out that the set with cardinal number $2^{\aleph_0}$ seemed to be much richer than any of the sets with the alephs $\aleph_1$, $\aleph_2$,... as their cardinal numbers which were constructed by an entirely different method. The investigation of the continuum hypothesis, which was also Hilbert's first problem, thus became an unproductive and rather pointless pursuit.

Disappointment seems to have coloured the attempt at the

solutions of many important problems of the nineteenth century, but one problem which troubled mathematicians for many years had a happier ending. In 1852 Francis Guthrie suggested that it was possible to colour any map, such as could be found in an atlas, so that no two adjacent countries or regions had the same colour using only four colours. Via his brother he passed the problem to Augustus De Morgan, another famous British nineteenth century mathematician. De Morgan passed it to Arthur Cayley who, in the first paper on the subject, reported that he could show the conjecture neither true nor false. This problem has continued to puzzle mathematicians up to the present, but no longer. In 1976 Kenneth Appel and Wolfgang Haken at the University of Illinois proved that four colours were sufficient to colour any plan or map. One aspect of the proof was apt to dishearten many mathematicians, it was achieved by reducing all the possible maps to a number of equivalent types which were then evaluated using several large electronic computers. Over 1,000 hours of computer time was needed! The normal methods had failed. No doubt mathematicians will still look for short, powerful and elegant methods but, as Appel and Haken have pointed out, there may be limits to what can be achieved by traditional means. There may be many other problems which will surrender under the weight of computer time where they would not to the more traditional methods.

As we have read, during the later part of the last century the mathematics department at the University of Göttingen was one of the leading centres of mathematical research. This continued into the twentieth century when the new Mathematics Institute was built, but was upset shortly after Hitler came to power in 1933. In that year Richard Courant, who had taken over Klein's old position, left Göttingen and took up a position, at first only temporary, at New York University. He took with him the manuscript of the uncompleted second volume of the Courant-Hilbert *Methods of Mathematical Physics*. This volume, which deals with partial differential equations, was published in German in 1937. It was an immediate success. Later it became useful to the war effort in both America and Germany and special editions were published in both countries for this purpose. Eventually an English translation was printed, which is still in use today.

In New York Courant set about founding and developing a new institute for mathematical excellence. With his characteristic skill and determination in these matters he succeeded in achieving his ambition on a grand scale and in 1964 the Courant Institute for Mathematical Sciences was opened in New York. Some, at least, of the traditions of Göttingen crossed the Atlantic.

All the mathematicians described in this book have contributed towards the present state of the subject and how it is interpreted and taught to all levels. Names and theorems do not have to be brought to mind to confirm their influence. They are not the only ones who have influenced our knowledge of mathematics, there is a long history, from before the ancient Greeks, which has given us this rich field of knowledge and activity.

Kurt Gödel (1906-1978) *by courtesy of the Institute for Advanced Study, Princeton, New Jersey*

Further Reading

General Histories of Mathematics
Carl B. Boyer, "A History of Mathematics", Wiley, New York, 1968.
Morris Kline, "Mathematical Thought from Ancient to Modern Times", Oxford University Press, New York, 1972.
Both of these are very extensive histories; Kline is the more mathematical of the two, Boyer is the more biographical.

Collections of Biographies of Mathematicians
E. T. Bell, "Men of Mathematics", Penguin Books, Harmondsworth, 1953 (originally published in 1937 in the U.S.A.).
Alexander MacFarlane, "Ten British Mathematicians", Wiley and Sons, New York, 1917.
Both of these include very interesting biographies of a number of the mathematicians in the present book. MacFarlane is now a fairly difficult book to find.

Biographies and Works of Individual Mathematicians
Galois
Leopold Infeld, "Whom the Gods Love: The Story of Evariste Galois", McGraw-Hill, New York, 1948.
Dedekind
Richard Dedekind, "Essays on the Theory of Numbers", Dover (reprint), New York, 1963.
Cantor
Georg Cantor, "Contributions to the Founding of the Theory of Transfinite Numbers", Dover (reprint), New York, 1955.
J. W. Dauben, "Georg Cantor", Harvard University Press, Cambridge (Massachusetts), 1979.

Peano

Hubert C. Kennedy, "Selected Works of Giuseppe Peano", Allen and Unwin, London 1973.

Hubert C. Kennedy, "Peano", D. Reidel, Dordrecht, Holland, 1980.

Hilbert

Constance Reid, "Hilbert", Springer-Verlag, New York, 1976.

These books are written at different levels; those by Infeld (this may be very difficult to find), Reid and Kennedy (1980) do not require a high level of mathematical knowledge. The others are mathematical but Dauben and Kennedy (1973) contain biographical sections which are well worth reading by those who are not looking for further mathematics.

Ancient Greek Mathematics

Sir Thomas L. Heath, "Greek Mathematics", Dover, New York, 1963. (This is a reprint of "A Manual of Greek Mathematics", Oxford University Press, 1931.)

Appendix: Group Axioms

A group is made up of a set $\{a, b, c....\}$ and an operation $*$ for combining the elements of the set, such that:
1) the combination of any two elements produces a further element of the set $a * b = c$, this is known as closure;
2) the order of combining any two adjacent elements is unimportant $(a * b) * c = a * (b * c)$, this is known as associativity;
3) there exists an identity element e so that $e * a = a$;
4) there exists inverse elements of the form a^{-1} etc., so that $a^{-1} * a = e$.

If also the operation is commutative

$a * b = b * a,$

then the group is said to be Abelian, after Niels Henrik Abel.

Subgroups are subsets of $\{a, b, c,...\}$ together with the operation $*$ such that the four group axioms are satisfied.

Index

Abel, Niels Henrik, 4, 11, 23, 93, 140
Abelian group, 105
Airy, G. B., 19
alephs, 77-79, 135
algebraic
  couples, 20, 21, 22
  integers, 66
  invariants, 51, 117
  numbers, 66
  structures, 129
Amis du Peuple, 8
'And', 43
Appel, Kenneth, 136
Aristotle, 31, 58, 73;
  'Physics', 73
Austin, Jane, 47
axioms of arithmetic, 120

Babylonians, 57
Bacon, Francis, 81
ballistics, 128
Bastille Day, 9
Berlin (university), 72, 73, 80, 85, 86, 116, 118, 121
Bernays, Paul, 123, 124
Berthod, M., 2
Board of Education (England) report on mathematics, 95
Bolyai, Janos, 83, 89, 122
Bolyai Prize, 122
Boole, George, 29-44, 50, 51, 102, 116
Boolean algebra, 29, 37-38, 41-44
Born, Max, 122, 126
'Bourbaki, Nicholas', 127-133
Brindley, Dr. 17, 18
British Association for the Advancement of Science, 22, 26, 50, 51
Brougham Bridge, 23
Brouwer, Jan, 123, 130
Brunsviga calculating machine, 95

Cantor-Dedekind axiom, 62
Cantor, George, 65, 68, 69-82, 103, 134, 135
cardinal number, 78, 79, 120
cat, 107
Cauchy, Augustin-Louis, 5, 11, 74
Cayley, Arthur, 33, 45-54, 93, 117, 118, 136
Cayley diagram, 54
Cayley-Hamilton theorem, 26
Cayley table, 54, 96
Chevalier, Auguste, 7, 10, 11, 12, 13, 16
Chevalley, Claude, 128
circuit, 29, 42-44

INDEX

Cohen, Paul, 135
Colburn, Zeah, 16
complex variable, 92
composition series, 13
computer, 43, 44, 136
conical refraction, 19-20
continuum hypothesis, 79, 80, 120, 135
continued fractions, 5
Courant-Hilbert, "Methods of Mathematical Physics", 124, 136
Courant Institute for mathematical Sciences, New York, 137
Courant, Richard, 97, 123, 124, 126, 128, 136
cube, 91, 92

Dedekind, cut, 64
Dedekind, Richard, 55-68, 138
Dehn, Max, 120
Delsante, Jean, 128
De Morgan, Augustus, 31, 36, 37, 136
De Morgan laws, 41, 44
determinants, 116
d'Herbinville, Pecheux, 10
diagonal proof (real numbers), 77
Dieudonne, Jean, 128
Dinet, M., 6
Dirichlet, Peter Custav Lejeune-, 61, 65
dodecahedron, 91, 92
D'Ovidio, Enrico, 100
duel, 10
Dumas, Alexandre, 8, 10

e, 63
Ecole Normale, 7, 8, 127
Ecole Polytechnique, 4, 5
Egyptians, 57
eigenfunctions, 121
eigenvalues, 121

Eilenberg, Samuel, 131
Einstein, Albert, 87, 123, 135
elliptic functions, 8, 61, 120;
    modular, 92
equations,
    cubic, 3, 11, 12, 13
    differential, 34-36, 116, 122
    integral, 121
    linear, 52
    polynomial, 65-66
    quadratic, 3, 11, 65, 66
    quintic, 4, 11, 12, 13, 15, 22, 91, 92
    seventh degree, 120
    sixth degree, 120
Erlanger Programme, 86-89, 90
Esperanto, 110
Euclid, 65, 83;
    postulates, 88-89
Euler, Leonhard, 6
Eulerian integrals, 61

field, 67
fifth postulate (Euclid), 88, 90
Formalist School, 122
Forsyth, A. R., 39
four colour conjecture, 136
Fourier, Joseph, 7, 11, 73
fractions, 55
Frege, Gottlob, 102
Fuchs, Lazarus, 116

Galileo, 74, 75, 76, 77
Galois, Evariste, 1-13, 15, 22, 23, 53, 61, 83, 86, 93, 116, 138
Galois fields, 11;
    groups, 11, 61
    theory, 11, 61
Gauss, Carl Frederick, 10, 60, 61, 65, 72, 74, 83, 89
Genocchi, Angelo, 101
geometry,
    afine, 88, 90
    co-ordinate, 83

Geometry *continued*
  Euclidean, 83, 90, 119, 130
  Greek, 83, 89
  hyperbolic, parabolic, elliptic, 89-90
  line, 85
  non-Euclidean, 88, 89, 90, 119
  point, 85
  projective, 87, 89, 90
Gibbs, Josiah Willard, 25
Gödel, Kurt, 135
Gordon, Paul, 90; (problem) 117-118
Göttingen (university), 60, 61, 62, 72, 80, 83, 85, 86, 93, 94, 97, 115, 116, 118, 121, 122, 123, 124, 125, 126, 128, 136, 137
Grassman, Hermann Gunther, 25
Greeks, ancient, 57, 59, 63, 65, 69, 82, 89, 137, 139
Gregory, Duncan, F., 33, 34
Gregory, James, 33
group, 11, 12, 13, 15, 28, 53, 54, 61, 86-88, 92, 95, 96, 105, 129, 140
Guthrie, Francis, 136

Haken, Wolfgang, 136
Hamilton, Sir William, 36
Hamilton, Sir William Rowan, 15-28, 32, 36, 40, 48, 50, 130
Hamiltonian graphs, 28; groups, 28
H.C.F. (highest common factor), 67
Heaviside, Oliver, 25
Heine, Eduard, 65
Helmholtz, Hermann v., 116
Hermite, Charles, 63, 117
Hilbert, David, 33, 66, 79, 103-4, 107, 108, 115-126, 127, 130, 134, 139
Hilbert space, 122, 124
Huntingdon, Edward, V., 41
Hurwitz, Adolf, 116, 117

icosahedron, 91
ideal, 67
ideal numbers, 66
incommensurable numbers, 59
incompleteness theorem (Gödel), 135
infinite (transfinite) cardinal numbers, 77-79, 120, 135
infinite sets, 77-79, 120
infinity, 73, 74
integers, 55, 57, 63
International Congress of Mathematics, Paris (1900), 107, 119;
  Rome (1908), 94;
  Cambridge (1912), 110;
  Bologna (1928), 125
International Congress of Philosophy, Paris (1900), 107
International Commission on Mathematical Instruction, 94
Intuitionist School, 123, 125
irrational numbers, 58, 59, 63, 64

Jacobi, Carl Gustav Jacob, 10, 22, 61, 116
Jevons, W. S. 41
Jordan, Camille "Traite des Substitutions et des equations algebrique", 11, 86

Kant, Immanuel, 115
Klein bottle, 96, 97
Klein, Christian Felix, 11, 83-97, 115, 117, 118, 123, 127, 136
Klein Four Group, 95-96
Kronecker, Leopold, 65, 72, 76, 79, 80, 99, 116, 121
Kummer, Ernst Eduard, 66, 72, 116

Lagrange, Joseph-Louis, 21, 22, 32

Laplace, Pierre-Simon de, 16, 17
Latino sine flexione, 106, 110, 111
least action, principle of, 21, 22
Leibniz, Gottfried Wilhelm, 31, 106
Lie, Sophus, 86
limits, 74, 75
Lindemann, Ferdinand, 63, 117, 118
Liouville, Joseph, 10, 63
Littlewood, J. E., 128
Lobatchevsky, Nicolai Ivanovich, 83, 89
logarithms, 6, 111
logic, 29, 31, 36, 38

Mathesis, 11
matrices, 45, 52, 53, 134
mechanics, 19, 21
Mendelssohn, Felix, 61
Meraner Lehrplanetwurfe, 94
Meray, Charles, 65
Minkowski, Hermann, 115, 117, 121, 122, 123
Mittag-Leffler, Gösta, 79, 80, 81
Möbius band, 96, 97

n-dimensional geometry, 51
nabla, 25
Nand, 44
Newton, Sir Isaac, 15, 26, 33, 128
Noether, Emmy, 126
nomography, 120
Nor, 44
'not', 43
Numbers,
  algebraic, 66
  complex, 15, 20-21, 22, 23, 130
  ideal, 66
  incommensurable, 59
  irrational, 58, 59, 63, 64, 65
  integers, 55, 57, 63
  natural, 55, 63, 74, 76, 77, 102
  prime, 11, 67

Numbers *continued*
  rational, 55, 57, 59, 63, 66
  real, 62-63, 77, 129
  transcendental, 63
  transfinite cardinal, 77-79
  transfinite ordinal, 78

octahedron, 91, 92
optics, 15, 18-19, 21
'or', 43

Padoa, 108
parallel postulate, 88, 90
parallelism, 88
parallelogram of forces, 16-17
Peano, Giuseppe, 76, 99-113, 126, 127, 134, 139
Peano's axioms (postulates), 102, 135
permutations, 11, 12, 13, 53, 54, 61
pi, π, 63
Pierce, C. S., 41, 102
Plato, 91, 'Timaeus', 91
Platonic solids, 91
Plücker, Julius, 85
Poincaré, Henri, 117, 122, 129
Poisson, Simeon-Denis, 8, 32
polyhedra, 120
power set, 78
Prandtl, Ludwig, 94
prime number, 11, 67
probability, 38
Pythagoras, 57, 58, 128; theorem, 57
Pythagorean triples, 57
Pythagoreans, 58

quaternions, 15, 23, 24-25, 26, 48, 53
quantum theory, 123